Sourcebook for Chemistry and Physics

SOURCEBOOK for CHEMISTRY and PHYSICS

David R. Hittle
Bay de Noc Community College

Frank D. Stekel
University of Wisconsin,
Whitewater

Shirley L. Stekel
University of Wisconsin,
Whitewater

Hans O. Andersen
Indiana University

Macmillan Publishing Co., Inc.
New York
Collier Macmillan Publishers
London

Copyright © 1973, Macmillan Publishing Co., Inc.
Printed in the United States of America

All rights reserved. No part of this book may be reproduced or
transmitted in any form or by any means, electronic or mechanical,
including photocopying, recording, or any information storage and
retrieval system, without permission in writing from the Publisher.

Macmillan Publishing Co., Inc.
866 Third Avenue, New York, New York 10022

Collier-Macmillan Canada, Ltd., Toronto, Ontario

Library of Congress Cataloging in Publication Data

Sourcebook for chemistry and physics.

   Bibliography: p.
   1. Chemistry—Study and teaching (Secondary)
2. Physics—Study and teaching (Secondary)
I. Hittle, David R.
QD40.S68   530'.07'12   72-90544
ISBN 0-02-354780-4

Printing: 1 2 3 4 5 6 7 8   Year: 3 4 5 6 7 8 9

Preface

A resource book can never be complete when it is designed to encompass all of the available resources for the chemistry and physics teacher. A number of compromises were necessary to limit the length of the book. Omissions are unintentional and frequently the assumption was made that the reader would have other references available. It is hoped that this book will be a useful source of information for the prospective teacher and the teacher in the classroom.

Part 1 is intended to provoke definitive thinking by the reader. Every individual views the scientific method in a slightly different way. The presentation in Part 1 should enable the reader to crystallize his own definition.

Parts 2 and 3 provide ideas and resources that may be used in the formulation of lesson plans. They are not complete with respect to either resources available or totality of concepts that occur in either chemistry or physics. These parts serve as a guide for the types of materials that need to be readily available when instruction is being planned.

Part 4 is intended to help the reader with specific problem areas that are unique to the chemistry and physics field. This part is not intended to supplant a regular methods text. The reader is encouraged to examine *Toward More Effective Science Instruction in Secondary Education* by Hans O. Andersen and Paul G. Koutnik, which is designed specifically for the problems of methodology. The examples used in these chapters are taken from our notes and should assure some degree of success.

The appendixes compile the information from a variety of sources we felt would be helpful.

We are specifically indebted to Vonda Hittle for reading the manuscript and offering suggestions that we feel make the book more readable. We are also indebted to the people with whom we work who have contributed ideas and encouragement. Last but not least, to the typists who struggled through the manuscript, we say, "Thank you."

D. R. H., F. D. S., S. L. S., H. O. A.

Contents

PART...1: **Introduction** 1

PART...2: **Resources for Teaching Chemistry** 5

    Aids for Chemical Concepts 7
        Scientific Investigation 7
        Atomic Structure 8
        Electrons 9
        Covalent Bonding 10
        Ionic Bonding 12
        Change of State 13
        Solids and Crystals 14
        Molecular Geometry 15
        Molecular Orbitals 17
        Gases 18
        Stoichiometry 19
        Solutions 21
        Colloids 23
        Acids and Bases 24
        Equilibria 25
        Electrochemistry: Oxidation-Reduction Reactions 27
        Reaction Mechanisms and Kinetics 28
        Thermodynamics 29
        Periodic Properties of the Elements 31
        Elements: Family IA and IIA 33
        Elements: Family IIIA and IVA 34
        Elements: Family VA and VIA 35
        Elements: Family VII and VIII 37
        Elements: Transition 38
        Nuclear Chemistry 39
        Coordination Chemistry 41
        Organic Chemistry 42
        Biochemistry 43
        Environmental Chemistry 45
        Spectral Analysis 46

PART...3: Resources for Teaching Physics 49

    Performance Objectives 50
      Distance Measurement by Triangulation 50
      Average Velocity and Instantaneous Velocity 52
      Velocity and Acceleration in Linear Motion 53
      Inertia 55
      Force = Mass × Acceleration 57
      Gravitational Fields and Apparent Weightlessness 58
      Newton's Third Law of Motion 61
      Motion Through Space 63
      Vectors 65
      Conservation of Momentum 67
      Energy Transformations and Conservation of Energy 68
      Conservation of Energy and Projectile Motion 70
      Uniform Circular Motion 72
      Temperature 73
      Heat and Work 75
      Wave Motion 78
      Reflection 79
      Refraction 81
      Interference 83
      Electrostatic Interactions 85
      Lines of Force of Electrostatic and Magnetic Fields 86
      Effect of a Magnetic Field upon Moving Charged Particles 88
      Radioactive Decay 91
      Contemporary Physics 92

PART...4: Additional Resources for the Teacher of Chemistry and Physics 95

    Performance Objectives 96
    Lesson Planning 109
    Lesson Plans 115
    Textbooks and Programs 138
    Utilizing Instructional Media 166
    Free and Inexpensive Materials 197
    The Functional Laboratory 204
    Laboratory Safety 226

Science Fairs 253
Periodicals and Organizations 260
Library Resources 271

PART...5: Appendixes 293

- A. General Laboratory Equipment 294
- B. Student Equipment 296
- C. Experiment Requirements Sheets 297
- D. Chemicals Used Frequently in the High School Chemistry Laboratory 300
- E. Inventory Card for Chemicals and Equipment 304
- F. Solution Preparation Formulas 305
- G. Student Safety Rules for the Chemistry Laboratory 312
- H. Accident Record Form 313
- I. Checklist for Shutting Down the Laboratory 314
- J. Sources of Chemistry and Physics Equipment and Supplies 315
- K. Directory of Educational Publishers 319
- L. Sources of Motion Pictures and Slidefilms—Free Loan 322
- M. Sources of 8 and 16mm Motion Picture and Slidefilm Libraries—Sales/Rentals 339

PART...1

Introduction

Sixteen years of education are now behind you, and many years of teaching chemistry and physics lie ahead. In the preparation of lesson plans for use in the classroom, the teacher must consider many questions. What is the chemistry or physics that must be taught to the students? What is the scientific method that seems to be an integral part of these areas of study? What assumptions can be made about the prospective students of chemistry or physics? What limitations must be assumed in relation to the teacher's capabilities to teach and to motivate students? What methods and resources are available to the teacher to increase his effectiveness? How much material must be "covered"? What methods of evaluation will be reliable and comprehensive?

An examination of the credentials of most teachers of chemistry or physics will show many courses taken in those departments. The teacher has been engaged in intensive study of specific parts of each discipline and often fails to look at the scope of the subject he has studied. The objectives of the courses that he has taken often have not been stated and the teacher has

2 Introduction

not had to rationalize the usefulness of studying all of the concepts that were presented. The teacher will frequently lack an explanation for the broad objectives of his discipline or the purpose that study in this area will serve.

The science that has been studied is in fact a very long collection of knowledge about the universe. As it is studied, the inevitable conclusion would be that science is ordered and that the communication of this order to high school students should be a prime objective. Many facts have been learned that have been useful to the teacher, and the teacher's tendency will be to pass these facts on without consideration for how the students will respond to them and use them. The teacher must first define science as he intends to teach it. That definition must be kept in mind as he looks at the other components of the teaching-learning process.

A definition of science that can be extended to either chemistry or physics is: (a) a collection of facts about the universe, (b) information that may be used to make predictions, (c) a synthesis of large quantities of data reduced to simpler theories, or (d) an explanation of why matter and energy behave as they do. This particular concept of science dwells on information and the theories that this information has produced. It says nothing about the information gathering process or what limits must be placed on the study of science.

The scientific method that has been expounded as the answer to all problems does have limitations that should be noted. The temptation is great to list the steps that are most frequently used to solve problems and to teach them as absolutes. Careful thought should reveal that the problem-solving procedure actually used varies with the individual, and even within the individual it will vary from problem to problem. What should be emphasized is that the information-gathering and problem-solving process is systematic. It is only one step further to conclude that science is a process and that the accumulated wisdom in the many books is merely a record of someone else's attempt to solve problems of that time. For the student to develop an appreciation of science should require that he develop his own logical process for acquiring information and solving problems that are relevant to his environment.

Having decided that the student will need both information and a process for adding new data to his working cognitive structure, the teacher needs to know what capabilities the student possesses when he enters the class. One solution to this problem is to administer a battery of tests at the beginning of school, and from this information, the quantity of new information to be taught can be determined. This assumes that the student needs the selected information to be a functional individual. If the student has no desire to retain a given piece of information, valuable instructional time will have been lost. It is generally accepted that it is impossible for any student of chemistry or physics to learn and retain the information presently found in most outlines. If the emphasis is shifted from the presentation of new information to acquisition procedures, the student should respond more readily. The response can be enhanced by selecting those problem areas that demonstrate the principles that provide the structure for the course. The student should also view the problem as being realistic and worthy of solution. Students also respond when they are actively involved. The inquiry approach has met with success because of the opportunity it affords the student to interact with his environment, which also includes the teacher.

In determining how to involve the students with the study of physics and chemistry, the teacher must assess his own strengths and weaknesses. If the teacher desires to present his knowledge in a concise fashion, as a finished product, he will typically not be successful in eliciting student response. Some inexperienced instructors cannot involve the students in successful group discussions or strategy sessions. The teacher must honestly assess his own willingness to try a particular strategy; for a positive attitude of the teacher will contribute to success in group discussion situations. The teacher must also be willing to learn from unsatisfactory situations. If the students fail to achieve some objectives, the teacher must question his presentation and the students' opportunities in order to determine where improvement could be made.

The limitations on teaching methods and resources are only bounded by the teacher's imagination. This book lists many free sources of virtually every kind of material for interesting the students and retaining their attention. If the budget is limited, the teacher must improvise and not use budget requirements as an excuse to return to a lecture presentation. The community can provide the raw materials, and the student will frequently assist in assembling components into finished products.

There will be a temptation to use lesson plans again. This must be done with great caution. If filed notes are used without revision, mistakes will probably be repeated that could have been eliminated. The teacher must be willing to depart from the prescribed sequence when circumstances warrant it. Go with the students. If they want an explanation to a problem, use that opportunity to present the relevant concepts because at this point they are interested.

Don't be trapped by a set content outline. In larger departments, the amount of material to cover is frequently specified and so may the time period. It is safe to say that such outlines have been in the files for some time and their author is more than likely retired.

Rigid content outlines will probably also reflect the emphasis on content that was prevalent before the curriculum projects were written. You should obtain permission from the department chairman and revise objectives to suit the needs of your student as dictated by data from tests and personal observations. When a teacher departs from the mean, he should be prepared and able to defend his actions. A good set of lesson plans and a daily log of the class activities including qualitative evaluations and notes for revision should be kept. This may be used at a later date to persuade other teachers that new methodologies do work, if tried.

Even when the teacher writes his own objectives, he must be willing to make modifications as he proceeds. The ability and interest of the students vary from year to year and from class section to class section. The teacher should stretch out a particular concept if the students are failing to understand it or if they are so excited about it that they want to explore it in greater depth. It is advisable to let the student determine as much as possible when he wants new study material.

Student evaluation is frequently the low point of any course. Teachers absorb themselves in the project of "giving grades". It seems that the purpose of instruction suddenly becomes a game in which those who know the rules can win the best prizes without respect to their actual abilities or knowledge. Tests are viewed as something to be endured by both the student and the teacher. Ideally, the test would be just another means of instruction and the teacher would anticipate the scores as reflecting positively on his ability to arrange the learning environment to insure success. The problems related to grading are great and cannot be satisfactorily handled in just a few words. The teacher is encouraged to face the problem and to propose radical solutions without respect to tradition. The scientific method demands that improvement in the system continue. The teacher should keep in mind as long as possible the feeling of disgust that may be directed toward a teacher who has not done an adequate job of evaluation. Remember the arbitrary way that claims for credit for alternate solutions were handled, and always remember what it was like to be a student. Do as much as you can to bring about change before the system takes over again.

It is expected that teachers will be vastly changing in the years ahead. If the textbook no longer determines the course content, then the chemistry or physics laboratory can become a fascinating place for a student. This book is intended to provide the chemistry and physics teacher with sufficient ideas and resources to begin the learning-teaching revolution in science classrooms.

PART...2

Resources for Teaching Chemistry

The reward of teaching chemistry or physics is reflected in the achievements of the students. To that end, this unit has been organized to aid the teacher in assembling the materials that will assist the student in mastering the selected concepts. For the beginning teacher, this unit should be a point of departure. The demonstrations that are listed are only a small portion of the large number that are available. References will be supplanted by newer interpretations, and newer films will help the student conceptualize far better than could any lecture or printed material. It is hoped that the teacher will become familiar with sources such as those used in this unit and that he will add to the materials presented here.

The objective given for each concept is described only in general terms and is to serve as a guide to the teacher when he writes his own. Only the teacher can fill in the situation and define the level of performance that he will require from his students. Many more specific objectives can be written for each concept. In the development of any unit of instruction that may include the concepts presented here, the teacher must add many more of his own objectives.

The demonstrations listed have been tried in the classroom by a large number of teachers over several years. Undoubtedly, the teacher is already familiar with many of them. Most of the demonstrations are completely safe; however, the instructor must always be ready for that one time when an error may occur. Notes of caution have been added where experience dictates the need for extra precautions. These demonstrations are easy to use and will help develop the habit of using demonstrations on a regular basis. However, demonstrations should not take the place of laboratory experience. The teacher is encouraged to examine some of the sources cited for demonstrations and add to the list.

The teacher and student references are generally separated for convenience. Where articles are listed, the student can usually profit by reading them as well as the teacher. Numerous appropriate articles are listed at the end of each chapter of *General Chemistry* by F. C. Schmidt, et al. Many other articles will be published in *Chemistry, Physics Today, Journal of Chemical Education, The Physics Teacher, Scientific American,* and other magazines. The teacher should add these articles to the reading list as they become available. Revised copies of the texts cited will also be published. These should be acquired and used as they become available.

The use of films requires both skill and luck to be effective. The 8mm films now being produced are inexpensive enough to be purchased by the school for repeated use. Many of these films will become available in the next few years. The list of 16mm films is not complete; the teacher must update this list as new films are produced. The firms listed are those that sell films. If film rental is desired, a catalogue of the available films should be secured from one of the outlets in your home state. A list of these outlets may be found in the Appendix.

Aids for Chemical Concepts

Scientific Investigation

OBJECTIVE

The student shall be able to demonstrate an understanding of the scientific method by utilizing a logical sequence of activities to solve a problem posed by the teacher.

DEMONSTRATION

1. The ability to hypothesize and use the nonvisual senses in the collection of data is enhanced by using a series of "Black boxes." These boxes are taped shut with an object inside that only the teacher knows. The students in groups of two or four can work with a single box. The students should list the data they collect, the predictions they make, and the observations they make to confirm the hypothesis. The uncertainty of science can be emphasized by not letting them examine the objects visually.

2. Observing the lighted candle is now a standard method of introducing the students to expand their powers of observation. A list of fifty-three observations that have been made can be found in Parry, R. W., et al., *Chemistry: Experimental Foundations.* Englewood Cliffs, N.J.; Prentice-Hall, Inc., 1970.

STUDENT REFERENCES

Cotton, F. A., and L. D. Lynch, *Chemistry: An Investigative Approach* (Boston: Houghton Mifflin Company, 1968), Chapter 1.
Chemical Bond Approach Project, *Chemical Systems* (New York: McGraw-Hill Book Company, Inc., 1964), Chapter 1.
O'Conner, P. R., et al., *Chemistry: Experiments and Principles* (Lexington, Mass.: D. C. Heath & Company, 1968), Chapter 1.
Parry, R. W., et al., *Chemistry: Experimental Foundations* (Englewood Cliffs, N.J.: Prentice-Hall, Inc., 1970), Chapter 1.

TEACHER REFERENCES

Cotton, F. A., and L. D. Lynch, *Teacher's Guide to Chemistry: An Investigative Approach* (Boston: Houghton Mifflin Company, 1968), Chapter 1. (Any introductory science book or any science methods book will deal with the scientific methods.)

8 Resources for Teaching Chemistry

16mm FILMS
Patterns of Scientific Investigation—Association Instruction Materials (color, sound, 22 min).
Random Events—Modern Learning Aids (b/w, sound, 31 min).
The Scientific Method—Encyclopedia Britannica Films (color, sound, 12 min).
Science and Foresight—Peter M. Roebeck & Co., (b/w, sound, 25 min).

8mm FILMS
Likely or Not? (Reasoning: Applied Concepts)—Encyclopedia Britannica Films (color, silent, 3 min).
Necessary But Not Sufficient (Reasoning)—Encyclopedia Britannica Films (color, silent, 3 min).

Atomic Structure

OBJECTIVE

The student shall be able to list the parts of an atom, state the properties of each, and cite appropriate experiments that confirm those properties.

DEMONSTRATION
1. Charge on an electron.
   Supplies: Pith ball electroscope, solid glass rod plus one polystyrene rod (-), a piece of fur and a piece of silk, polyethylene rod (-), sealing wax sticks (-).
   Procedure: Stroke each rod with both pieces of cloth and allow the students to hypothesize into which item (rod or material) the electrons are moving. Establish the fact that the rods are charged differently or alike by holding the subsequently charged rod close to the fully charged pith balls. Discharge the pith balls by touching them with the hand.
2. Deflection of an electron beam.
   Supplies: Magnetic effect tube (Frey Scientific, cat. no. 1200) or monitored induction coil (Frey Scientific, cat. no. 3183), bar magnet (Frey Scientific, cat. no. 267.)
   Procedure: Connect the induction coil to the magnetic effect tube, turn on the coil, and alternate the ends of the magnet. (This demonstration is most effective when the room is partially lighted).

STUDENT REFERENCES
Cotton, F. A., and L. D. Lynch, *Chemistry: An Investigative Approach* (Boston: Houghton Mifflin Company, 1968), Chapter 20.
Chemical Bond Approach Project, *Chemical Systems* (New York: McGraw-Hill Book Company, 1964), Chapter 7.

Kendall, H. W., and W. Panofsky, "The Structure of the Proton and the Neutron," *Scientific American* (June, 1970), p. 60.

O'Conner, P. R., et al., *Chemistry: Experiments and Principles* (Lexington, Mass.: D. C. Heath & Company, 1968), Chapter 8.

Parry, R. W., et al., *Chemistry: Experimental Foundations* (Englewood Cliffs, N.J.: Prentice-Hall, Inc., 1970), Chapter 7.

TEACHER REFERENCES

Cotton, F. A., and L. D. Lynch, *Teacher's Guide for Chemistry: An Investigative Approach* (Boston: Houghton Mifflin Company, 1968), Chapter 5.

Hecht, S., *Explaining the Atom* (New York: The Viking Press, Inc., 1964).

Lagowski, J. J., *The Structure of Atoms* (Boston: Houghton Mifflin Company, 1964).

Schmidt, F. A., et al., *General Chemistry* (Lexington, Mass.: D. C. Heath & Company, 1972), Chapter 3.

Steams, R. L., *Focus on Physics—Atomic Physics* (New York: Barnes & Noble, Inc., 1969).

16mm FILMS

Mass of the Electron—Modern Learning Aids (b/w, sound, 18 min).
Rutherford Atom—Modern Learning Aids (b/w, sound, 40 min).
Crooke's Tubes—Association Instructional Materials (color, silent, 8 min).

8mm FILMS

Rutherford Scattering—Harper & Row, Publishers (color, silent, 4 min).
Thomson Model of the Atom—Ealing Film Loops (color, silent, 4 min).
Particles of Matter—Universal Education and Visual Aids (color, silent, 12½ min).

Electrons

OBJECTIVE

The student shall be able to list the properties of electrons, including mass, charge, spin, order in which they fill orbitals, and the shapes of the orbitals they fill.

DEMONSTRATION

1. See demonstrations listed in Atomic Structure.

2. Obtain a set of s, p, d, f orbital models and have it available on the demonstration desk.

3. The uncertainty of the position and velocity of the electron may be illustrated by tossing a volleyball out to the students and plotting the path that the ball takes from student to student as it moves around the room. The path can be predicted in advance and followed in actuality to make the comparison, or the uncertainty can be followed by looking at the path traced out as the volleyball moves around the room.

STUDENT REFERENCES
Cotton, F. A., and L. D. Lynch, *Chemistry: An Investigative Approach* (Boston: Houghton Mifflin Company, 1968), Chapter 8.
Chemical Bond Approach Project, *Chemical Systems* (New York: McGraw-Hill Book Company, Inc., 1964), Chapter 7.
O'Conner, P. R., et al., *Chemistry: Experiments and Principles.* (Lexington, Mass.: D. C. Heath & Company, 1968), Chapter 9.
Parry, R. W., et al., *Chemistry: Experimental Foundations* (Englewood Cliffs, N.J.: Prentice-Hall, Inc., 1970), Chapter 7.

TEACHER REFERENCES
Bordass, W. T., and J. W. Linnett, "A New Way of Presenting Atomic Orbitals," *Journal of Chemical Education*, 47:672 (1970).
Cotton, F. A., and L. D. Lynch, *Teacher's Guide to Chemistry: An Investigative Approach* (Boston: Houghton Mifflin Company, 1968), Chapter 8.
Hochstrasser, W. A., *Behavior of Electrons in Atoms* (Menlo Park, Calif.: W. A. Benjamin, Inc., 1962).
McClellan, A. L., "Transparent 3-D Models of Electron Probability Distributions," *Journal of Chemical Education*, 45:761 (1970).
Schmidt, F. C., et al., *General Chemistry* (Lexington, Mass.: D. C. Heath & Company, 1972), Chapter 3.
Sisler, H. H., *Electronic Structure, Properties, and the Periodic Law* (New York: Van Nostrand Reinhold Company, 1963).

16mm FILMS
The Hydrogen Atom—Modern Learning Aids (color, sound, 20 min).
The Principle of Uncertainty—Peter M. Roebeck and Co. (b/w, sound, 30 min).
Electrons at Work—Encyclopedia Britannica (b/w, sound, 15 min).
E/M Demonstration—Encyclopedia Britannica (color, silent, 4 min).
Crookes Tubes—Encyclopedia Britannica (color, silent, 9 min).

8mm FILMS
Oil Drop Experiment (Millikan's)—Encyclopedia Britannica (b/w, silent, 7 min).

Covalent Bonding

OBJECTIVE
The student shall be able to define covalent bonding in terms of the sharing of electrons and also to differentiate between sigma and pi bonds, with respect to relative bond strength.

DEMONSTRATION
1. See the demonstrations listed under Ionic Bonding.
2. The concept of a double bond may be demonstrated by comparing the results of the addition of bromine to a saturated and an

unsaturated hydrocarbon. A few drops of bromine should be placed in a large flask and stoppered. Upon the addition of a milliliter of the appropriate hydrocarbon, the brownish vapor will disappear when it is shaken with the unsaturated hydrocarbon. Examination of structural formulas should convince the students that one hypothesis is the presence of a double bond.

3. Alyea, H. N., and F. B. Dutton, *Tested Demonstrations in Chemistry*, Easton, Pennsylvania, Division of Chemical Education of the American Chemical Society, 1965, pp. 36, 37, 39, 42, 47, 48, and 107-113. This source lists demonstrations that may be used to emphasize the properties of covalently bonded compounds.

STUDENT REFERENCES

Cotton, F. A., and L. D. Lynch, *Chemistry: An Investigative Approach* (Boston: Houghton Mifflin Company, 1968), Chapter 9.

Chemical Bond Approach Project, *Chemical Systems* (New York: McGraw-Hill Book Company, 1964), Chapters 9 and 10.

O'Conner, P. R., et al., *Chemistry: Experiments and Principles* (Lexington, Mass.: D. C. Heath & Company, 1968), Chapter 10.

Parry, R. W., et al., *Chemistry: Experimental Foundations* (Englewood Cliffs, N.J.: Prentice-Hall, Inc., 1970), Chapter 8.

TEACHER REFERENCES

Cotton, F. A., and L. D. Lynch, *Teacher's Guide to Chemistry: An Investigative Approach* (Boston: Houghton Mifflin Company, 1968), Chapter 9.

Griswold, Ernest, *Chemical Bonding and Structure* (Lexington, Mass.: D. C. Heath & Company, 1968).

Lynch, P. F., *Orbitals and Chemical Bonding* (Boston: Houghton Mifflin Company, 1969).

Margolis, E. J., *Bonding and Structure: A Review of Fundamental Chemistry* (New York: Appleton-Century-Crofts, 1968).

Schmidt, F. C., et al., *General Chemistry* (Lexington, Mass.: D. C. Heath & Company, 1972), Chapter 4.

16mm FILMS

Chemical Bonding—Modern Learning Aids (color, sound, 16 min).
Chemical Bonding and Atomic Structure—Coronet Films (b/w, sound, 16 min).

8mm FILMS

Atomic and Bonding Orbitals—Willard Grant Press, Inc. (color, silent, 4 min).
The Properties of a Covalently Bonded Molecule—Longmans, Green & Company (color, silent, 3 min).
Simple Molecular Orbitals—Harper & Row, Publishers (color, silent, 4 min).
Atoms to Molecules—McGraw-Hill Book Company (series) (color, silent, 4 min).

Ionic Bonding

OBJECTIVE

The student shall be able to define ionic bonding in terms of electron transfer and be able to predict which compounds would probably be ionic based on the individual element's ability to retain electrons.

DEMONSTRATION

1. Test different solutions for their ability to conduct electricity. The apparatus should include two electrodes for placement in the solution and a light bulb in the circuit to indicate the degree of conductivity. If all of the solutions are 1 molar, then the students should be able to conclude that not all bonding is 100 percent ionic or covalent.
2. A comparison of the effect of equal concentrations of salt and sugar on the freezing point of the solution will also lead to the conclusion that there must be ion formation in the solution.
3. Alyea and Dutton, *Tested Demonstrations in Chemistry*, pp. 74-76, 148, 163 and 164.

STUDENT REFERENCES

Cotton, F. A., and L. D. Lynch, *Chemistry: An Investigative Approach* (Boston: Houghton Mifflin Company, 1968), Chapter 9.
Chemical Bond Approach Project, *Chemical Systems* (New York: McGraw-Hill Book Company, Inc., 1964), Chapters 12 and 13.
O'Conner, P. R., et al., *Chemistry: Experiments and Principles* (Lexington, Mass.: D. C. Heath & Company, 1968), Chapter 10.
Parry, R. W., et al., *Chemistry: Experimental Foundations* (Englewood Cliffs, N.J.: Prentice-Hall, Inc., 1970), Chapter 8.

TEACHER REFERENCES

Cotton, F. A., and L. D. Lynch, *Teacher's Guide to Chemistry: An Investigative Approach* (New York: Houghton Mifflin Company, 1968), Chapter 9.
Holliday, L., "Early Views on Forces Between Atoms," *Scientific American* (May, 1970), p. 116.
Lagowskik, J. J., *The Chemical Bond* (New York: Houghton Mifflin Company, 1966).
Lewis, G. N., *Valence and the Structure of Atoms and Molecules* (New York: Dover Publications, Inc., 1966).
Sanderson, R. T., "Principles of Chemical Bonding," *Journal of Chemical Education*, 38:382 (1961).
Schmidt, F. C., et al., *General Chemistry* (Lexington, Mass.: D. C. Heath & Company, 1972), Chapter 4.
Stevens, B., *Atomic Structure and Valency* (New York: Barnes & Noble, Inc., 1967).

16mm FILMS
Ionization Energy—Modern Learning Aids (color, sound, 22 min).
Ionization—Coronet Films (color, sound, 18½ min).
Ionization and Ionic Equilibrium—Indiana University (color, sound, 15 min).

8mm FILMS
None available as of this writing.

Change of State

OBJECTIVE

The student shall be able to propose a model for the change of state substances that incorporates the theory of kinetic molecular motion. Any model would also describe the degree of order in each state.

DEMONSTRATION

1. The direct change of a solid into a gas can be demonstrated with dry ice.
2. The existence of a liquid-gas equilibrium may be illustrated with a bottle of bromine liquid. The space above the liquid will be filled with bromine gas that is visible to the student.
3. Breakaway models that allow the teacher to remove layers from the chunk and finally split the layers into individual spheres can demonstrate to some extent the different degrees of order among the three states.

STUDENT REFERENCES
Cotton, F. A., and L. D. Lynch, *Chemistry: An Investigative Approach* (Boston: Houghton Mifflin Company, 1968), Chapter 4.
Chemical Bond Approach Project, *Chemical Systems* (New York: McGraw-Hill Book Company, Inc., 1964), Chapter 1.
O'Conner, P. R., et al., *Chemistry: Experiments and Principles* (Lexington, Mass.: D. C. Heath & Co., 1968), Chapter 5.
Parry, R. W., et al., *Chemistry: Experimental Foundations* (Englewood Cliffs, N.J.: Prentice-Hall, Inc., 1970), Chapter 5.

TEACHER REFERENCES
Cotton, F. A., and L. D. Lynch, *Teacher's Guide to Chemistry: An Investigative Approach* (Boston: Houghton Mifflin Company, 1968), Chapter 4.
Schmidt, F. C., et al., *General Chemistry* (Lexington, Mass.: D. C. Heath & Company, 1972), Chapter 10.
Tabor, D., *Gases, Liquids, and Solids* (Baltimore: Penguin Books, Inc., 1969).

16mm FILMS
Explaining Matter: Molecules in Motion—Encyclopedia Britannica (color, sound, 11 min).

8mm FILMS
Physical and Chemical Changes: Phase Demonstration—Encyclopedia Britannica (color, silent, 4 min).
Molecular Motion in Condensed Phases—Modern Learning Aids (color, silent, 5 min).
Critical Temperature—Harper & Row Publishers (color, silent, 4½ min).
Boiling Point and Pressure—Harper & Row Publishers (color, silent, 3½ min).

Solids and Crystals

OBJECTIVE
The student shall be able to classify crystals according to the type of unit cell. The student shall be able to differentiate between crystals and noncrystals based on the order involved.

DEMONSTRATION
1. Construct three-dimensional models of each unit cell from styrofoam spheres. Different colors should be used to show the anions and cations. Cutaway models should also be available if the student has difficulty in visualizing the entire unit cell. Models of hexagonal and cubic closest packing can also be prepared from these materials. Transparencies may be used in conjunction with these models but not alone.
2. A comparison of the melting points of crystals and amorphous solids can be demonstrated with phenylacetic acid and paraffin. Place a few crystals of phenylacetic acid in a 20 ml beaker and a small piece of paraffin in another 20 ml beaker. Both beakers should be placed in a large petri dish that is resting on the stage of an overhead projector. Add hot water ($90°C$) to the petri dish and observe the melting process of the substance in each beaker. The range of temperatures at which each melts could be determined in the laboratory.

STUDENT REFERENCES
Cotton, F. A., and L. D. Lynch, *Chemistry: An Investigative Approach* (Boston: Houghton Mifflin Company, 1968), Chapter 10.
Chemical Bond Approach Project, *Chemical Systems* (New York: McGraw-Hill Book Company, Inc., 1964), Chapters 11 and 12.
O'Conner, P. R., et al., *Chemistry: Experiments and Principles* (Lexington, Mass.: D. C. Heath & Company, 1968), Chapter 5.
Parry, R. W., et al., *Chemistry: Experimental Foundations* (Englewood Cliffs, N.J.: Prentice-Hall, Inc., 1970), Chapter 5.

TEACHER REFERENCES
Cotton, F. A., and L. D. Lynch, *Teacher's Guide to Chemistry: An Investigative Approach* (Boston: Houghton Mifflin, 1968), Chapter 10.
Evans, R. C., *An Introduction to Crystal Chemistry* (New York: Cambridge University Press, 1964).
Galwey, A. K., *Chemistry of Solids* (New York: Barnes & Noble, Inc., 1967).
Holden, Alan, and Phylis Singer, *Crystals and Crystal Growing* (Garden City, N.Y.: Doubleday & Company, Inc., 1960).
Moore, W. J., *Seven Solid States: An Introduction to the Chemistry of Solids* (Menlo Park, Calif.: W. A. Benjamin, Inc., 1967).
Schmidt, F. C., et al., *General Chemistry* (Lexington, Mass.: D. C. Heath & Company, 1972), Chapter 10.

16mm FILMS
Crystals—Modern Learning Aids (color, sound, 25 min).
A Study of Crystals—Journal Films (color, sound, 17 min).
Glass—The Royal Institute of Chemistry (b/w, sound, 21 min).

8mm FILMS
Thermal Expansion of Solids—Ealing Film-Loops (color, silent, 4 min).
Most Solids Melt—International Communication Films (color, silent, 4 min).
Identifying Solids by Density—Harper & Row (color, silent, 4 min).
Close Packing of Spheres—Harper & Row Publishers (color, silent, 4 min).
Crystal Growth—Harper & Row, Publishers (color, silent, 4 min).
Crystals and X-Ray Diffraction—Modern Learning Aids (b/w, silent, 4 min).

Molecular Geometry

OBJECTIVE
The student shall be able to predict the shape and polarity of a molecule from its formula and be able to predict the relative boiling point of that compound according to its shape and polarity with respect to other given compounds.

DEMONSTRATION
1. Use models of various types of covalently bonded compounds, which range from polar to nonpolar compounds. Molecules to consider could include H_2O, CH_4, CO_2, HCl, CCl_4, CH_3OH, and C_6H_6. The boiling points of polar and nonpolar compounds can be placed on an acetate sheet and projected for class consideration. As a result of the examination of polarity and the boiling points, the concept of the hydrogen bond should evolve.

2. Actual boiling points of molecules of differing polarities can be demonstrated by placing 2 ml of pentane, diethyl ether (USE CAU-

TION), and 1-butanol in 10 ml beakers with a couple of boiling beads. A large petri dish of hot water can be placed on an overhead projector, and the beakers can be placed into the petri dish. The order in which they boil should be indicative of their boiling points. Remember to identify the compound in each beaker and its position in the dish so that the students can observe the correct order of boiling.

3. A transparency should be prepared that plots molecular weight versus boiling point for the series H_2O, H_2S, H_2Se, H_2Te, NH_3, PH_3, AsH_3, SbH_3; and the inert gases. The concept of hydrogen bonding can be supported from these data.

STUDENT REFERENCES
Cotton, F. A., and L. D. Lynch, *Chemistry: An Investigative Approach* (Boston: Houghton Mifflin Company, 1968), Chapter 10.
Chemical Bond Approach Project, *Chemical Systems* (New York: McGraw-Hill Book Company, 1964), Chapter 7.
O'Conner, P. R., et al., *Chemistry: Experiments and Principles* (Lexington, Mass.: D. C. Heath & Company, 1968), Chapters 10 and 17.
Parry, R. W., et al., *Chemistry: Experimental Foundations* (Englewood Cliffs, N.J.: Prentice-Hall, Inc., 1970), Chapter 16.

TEACHER REFERENCES
Cotton, F. A., and L. D. Lynch, *Teacher's Guide to Chemistry: An Investigative Approach* (Boston: Houghton Mifflin Company, 1968), Chapter 10.
Herz, Werner, *The Shape of Carbon Compounds* (Menlo Park, Calif.: W. A. Benjamin, Inc., 1963).
Jennings, K. R., *Molecular Structure* (New York: Barnes & Noble, Inc.), 1969.
Orville-Thomas, W. J., *Structure of Small Molecules* (New York: American Elsevier Publishing Co., Inc., 1966).
Schmidt, F. C., et al., *General Chemistry* (Lexington, Mass.: D. C. Heath & Company, 1972), Chapter 10.
Ryschkewitsch, G. E., *Chemical Bonding and the Geometry of Molecules* (New York: Van Nostrand Reinhold Company, 1963).

16mm FILMS
Shapes and Polarities of Molecules—Modern Learning Aids (color, sound, 18 min).
The Structure of Water—McGraw-Hill Book Company, Inc. (color, sound, 14 min).

8mm FILMS
The Structure of a Covalent Molecule—CCl_4—Longmans, Green and Company (color, silent, 3 min).

Molecular Orbitals

OBJECTIVE

The student shall be able to predict which atomic orbitals will form molecular orbitals for a given homonuclear diatomic molecule and then provide a distribution that will enable him to predict whether the molecule would be paramagnetic or diamagnetic.

DEMONSTRATION

1. Prepare a transparency with the atomic orbitals of two atoms along the edges and with the appropriate molecular orbitals in the center in the accepted locations (Figure 2-1).
2. Show the shapes of σ_s, σ_s^*, σ_{px}^*, π_{py}, and π_{py}^*.

REFERENCES

Cooper, W. F., G. A. Clark, and C. R. Hare, "A Simple, Quantitative Molecular Orbital Theory, *Journal of Chemical Education*, 48:247 (1971).

George, J. W., "Hybridization in the Description of Homonuclear Diatomic Molecules," *Journal of Chemical Education*, 42:152 (1965).

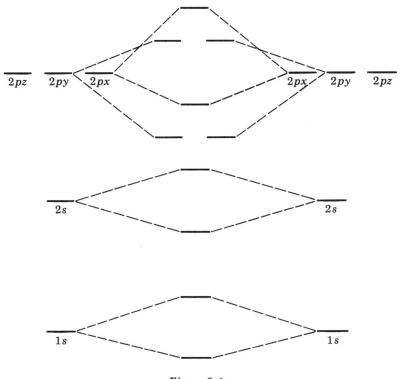

Figure 2-1

Gray, H., *Electrons and Chemical Bonding* (Menlo Park, Calif.: W. A. Benjamin, Inc., 1964).
Greenwood, N. N., "Chemical Bonds," *Education in Chemistry*, 4:164 (1967).
Griswold, E., *Chemical Bonding and Structure* (Lexington, Mass.: D. C. Heath & Company, 1968).
Hochstrasser, R. M., *Behavior of Electrons in Atoms* (Menlo Park, Calif.: W. A. Benjamin, Inc., 1964).
Schmidt, F. A., et al., *General Chemistry* (Lexington, Mass.: D. C. Heath & Company, 1972), Chapter 5.

8mm FILMS
Simple Molecular Orbitals—Harper & Row, Publishers (color, silent, 4 min).

Gases

OBJECTIVE
The student shall be able to apply the conventional gas laws in the determination of volume changes resulting from changes in the temperature, pressure, or number of moles of the gas.

DEMONSTRATION
1. A molecular motion demonstrator available from Educational Materials and Equipment Company may be used in conjunction with an overhead projector to demonstrate temperature effects on molecular motion as well as the process of osmosis and diffusion. Nine different properties of gases can be demonstrated with this equipment.
2. The pressure-volume relationship of a gas may be demonstrated by the use of a tire pump and a set of weights. With the outlet of the pump plugged, placing more weight on the handle will cause the piston to decrease the volume of the gas inside.
3. The temperature-volume relationship of a gas can be seen by inverting a flask having a piece of glass tubing inserted in a rubber stopper in its neck into a beaker of water. Pouring ice water over the flask will cause water to rise into the flask and the addition of hot water over the flask will cause the water level inside to decrease.
4. Diffusion can be illustrated by unstoppering various bottles of aromatic compounds located at different portions of the room. Time measurements can be made to determine the rate of diffusion of each compound tested.
5. Alyea and Dutton, *Tested Demonstrations in Chemistry*, pp. 63, 65, 156, and 160, details other demonstrations that may be used.

STUDENT REFERENCES

Cotton, F. A., and L. D. Lynch, *Chemistry: An Investigative Approach* (Boston: Houghton Mifflin Company, 1968), Chapter 3.

Chemical Bond Approach Project, *Chemical Systems* (New York: McGraw-Hill Book Company, 1964), Chapter 3.

O'Conner, P. R., et al., *Chemistry: Experiments and Principles* (Lexington, Mass.: D. C. Heath & Company, 1968), Chapter 4.

Parry, R. W., et al., *Chemistry: Experimental Foundations* (Englewood Cliffs, N.J.: Prentice-Hall, Inc., 1970), Chapter 4.

TEACHER REFERENCES

Cotton, F. A., and L. D. Lynch, *Teacher's Guide to Chemistry: An Investigative Approach* (Boston: Houghton Mifflin Company, 1968), Chapter 3.

Ferguson, J. L., "Liquid Crystals," *Scientific American* (August, 1964), p. 77.

Kauzmann, W., *Thermal Properties of Matter I: Kinetic Theory of Gases* (Menlo Park, Calif.: W. A. Benjamin, Inc., 1966).

Mendelson, K., *Quest for Absolute Zero* (New York: McGraw-Hill Book Company, Inc., 1967).

Parsonage, N. G., *The Gaseous State* (Elmsford, N.Y.: Pergamon Press Inc., 1970).

Schmidt, F. C., et al., *General Chemistry* (Lexington, Mass.: D. C. Heath & Company, 1972), Chapter 8.

Scott, Sir Neville, "The Solid State," *Scientific American* (September, 1967), p. 80.

Sime, R. J., "Some Models of Close Packing," *Journal of Chemical Education*, 40:54 (1963).

Updike, J., "The Dance of the Solids," (poetry), *Scientific American* (January, 1969), p. 130.

16mm FILMS

Change of State—The Royal Institute of Chemistry (color, sound, 20 min).
Behavior of Gases—Modern Learning Aids (b/w, sound, 15 min).
Gas Laws and Their Applications—Encyclopedia Britannica (b/w, sound, 13 min).

8mm FILMS

Absolute Zero—Modern Learning Aids (b/w, silent, 4 min).
Kinetic Theory—Harper & Row Publishers, set of 6. (color, silent, 4 min).
Molecular Motion—Harper & Row Publishers (color, silent, 4 min).

Stoichiometry

OBJECTIVE

The student shall be able to determine formulas of compounds if given the percentage of each element present. He shall also be able to calculate the quantity of a given product produced in a known chemical reaction if given the mass of the reactant that is completely consumed.

DEMONSTRATION

1. A previously weighed sample of HgO may be decomposed by heating strongly in a test tube. When the solid turns black, the weight of the remaining solid may be determined and the percentage of mercury and oxygen in the original sample and by using the standard mathematical procedures the formula can be determined. The teacher could also start by giving the students the formula and ask them to calculate how much Hg should be left in the test tube. Carrying out the heating of the sample will confirm the prediction.

2. A reaction of 500 ml of 0.1M $AgNO_3$ and 500 ml of 0.1M NaCl may be demonstrated in class. The teacher can explain how much of each reactant is present and can show how the mass of the resulting AgCl may be determined. To save time, a table may be constructed on the board listing the amount of $AgNO_3$ and NaCl in the reaction mixture. By using one of the procedures for calculating the mass of the product, the students can predict how much mass of AgCl should be present if each reactant is completely consumed. In this case the students should arrive at the same answer. The teacher should be asked to formulate hypotheses to explain any deviations from their predictions. The procedure, calculations, and results may be varied by reacting the $AgNO_3$ with a solution of $MgCl_2$.

3. Alyea and Dutton, *Tested Demonstrations in Chemistry*, pp. 6, 54, and 71-73 may be consulted to learn of other demonstrations that will broaden this concept.

STUDENT REFERENCES

Cotton, F. A., and L. D. Lynch, *Chemistry: An Investigative Approach* (Boston: Houghton Mifflin Company, 1968), Chapter 4.

Chemical Bond Approach Project, *Chemical Systems* (New York: McGraw-Hill Book Company, Inc., 1964), Chapter 3.

O'Conner, P. R., et al., *Chemistry: Experiments and Principles.* (Lexington, Mass.: D. C. Heath & Company, 1968), Chapter 3.

Parry, R. W., et al., *Chemistry: Experimental Foundations* (Englewood Cliffs, N.J.: Prentice-Hall, Inc., 1970), Chapter 3.

TEACHER REFERENCES

Cotton, F. A., and L. D. Lynch, *Teacher's Guide to Chemistry: An Investigative Approach* (Boston: Houghton Mifflin Company, 1968), Chapter 4.

Kieffer, W. F., *The Mole Concept in Chemistry* (New York: Van Nostrand Reinhold Company, 1962).

Margolis, E. J., *Formulation and Stoichiometry* (New York: Appleton-Century-Crofts, 1968).

Nash, L. K., *Stoichiometry* (Reading, Mass.: Addison-Wesley Publishing Co., Inc., 1966).

Schmidt, F. C., et al., *General Chemistry* (Lexington, Mass.: D. C. Heath & Company, 1972), Chapter 2.

16mm FILMS
Determining a Molecular Formula—McGraw-Hill Book Company, Inc. (b/w, sound, 12½ min).
Gases and How They Combine—Modern Learning Aids (color, sound, 22 min).
Synthesis of a Compound—Coronet Films (color, sound, 13½ min).

8mm FILMS
On Writing Equations (I)—Willard Grant Press (b/w, silent, 4 min).
On Writing Equations (II)—Willard Grant Press (b/w, silent, 4 min).

Solutions

OBJECTIVE

The student shall be able to demonstrate a knowledge of the solution process by describing the relationship of the solute molecules or ions to the solvent molecules. He shall also be able to handle the quantitative relationships of molarity, molality, and percent by weight in solution preparation and in determining the quantity of each that is necessary for a given problem.

DEMONSTRATION

1. The effect of temperature on the rate of solution can be demonstrated by having three large cylinders filled with water at three different temperatures sitting on the desk. Suggested temperatures are 20°, 40°, and 60°, but other temperature spreads may be used. Small crystals of potassium permanganate of approximately the same mass and size should be added to the cylinders.

2. The effect of temperature on solubility may be demonstrated by saturating a solution of sugar and water at room temperature. To the chosen volume of solution add 10 to 20 grams more sugar so the students can see that there is solid on the bottom. Heat and stir the solution until the remaining sugar dissolves. The student should be asked to formulate hypotheses that would explain the increased solubility.

3. Physical properties of the solvent and the solution for various solutes should be compared. The most common properties to compare are the boiling and freezing points. The regularity of the increase in the boiling point and the decrease in the freezing point can be plotted graphically for molecular solutes at concentrations of .1m, .2m, .3m, and so on. The students can be asked to extrapolate to a 1m solution. Data for various ionic solutes can be presented and the students should be able to derive the factor by which the boiling point elevation constant or the freezing point depression constant must be multiplied to obtain the actual solution temperatures.

4. Alyea and Dutton, *Tested Demonstrations in Chemistry*, pp. 66-68, 143, and 189 describe in detail other demonstrations that may be helpful to the teacher.

STUDENT REFERENCES
Cotton, F. A., and L. D. Lynch, *Chemistry: An Investigative Approach* (Boston: Houghton Mifflin Company, 1968), Chapter 10.
Chemical Bond Approach Project, *Chemical Systems* (New York: McGraw-Hill Book Company, Inc., 1964), Chapter 2 and 3.
O'Conner, P. R., et al., *Chemistry: Experiments and Principles* (Lexington, Mass.: D. C. Heath & Company, 1968), Chapter 6.
Parry, R. W., et al., *Chemistry: Experimental Foundations* (Englewood Cliffs, N.J.: Prentice-Hall, Inc., 1970), Chapter 5.

TEACHER REFERENCES
Cotton, F. A., and L. D. Lynch, *Teacher's Guide to Chemistry: An Investigative Approach* (New York: Houghton Mifflin Company, 1968), Chapter 10.
Dreisback, Dale, *Liquids and Solutions* (Boston: Houghton Mifflin Company, 1966).
Eisenberg, D., and W. Kauzmann, *The Structure and Properties of Water* (New York: Oxford University Press, 1969).
Gurney, R. W., *Ionic Processes in Solution* (New York: Dover Publications, Inc., 1962).
Hildebrand, J. H., and R. L. Scott, *The Solubility of Nonelectrolytes* (New York: Dover Publications, Inc., 1964).
Hunt, J. P., *Metal Ions in Aqueous Solution* (Menlo Park, Calif.: W. A. Benjamin, Inc., 1963).
Schmidt, F. C., et al., *General Chemistry* (Lexington, Mass.: D. C. Heath & Company, 1972), Chapter 12.
Zingaro, R. A., *Nonaqueous Solvents* (Lexington, Mass.: D. C. Heath & Company, 1968).

16mm FILMS
Solutions—Coronet Films (color, sound, 16 min).
Dynamics of Solution—McGraw-Hill Book Company, Inc. (color, sound, 14 min).
Liquids in Solution—McGraw-Hill Book Company, Inc. (color, sound, 11 min).
Properties of Solution—Coronet Films (color, sound, 28 min).

8mm FILMS
Identifying Liquids by Density—Harper & Row, Publishers (color, silent, 4 min).
Thermal Expansion of Liquids—Harper & Row, Publishers (color, silent, 4½ min).
Solutions—Sutherland Educational Films (color, silent, 3 min).
Identifying Fabrics by Solubility—Ealing Film Loops (color, silent, 4 min).

Colloids

OBJECTIVE

The student shall be able to differentiate between colloidal systems and solutions based on whether the substance dispersed is molecular or particular. He shall also be able to identify at least five forms of pollution that produce colloidal systems.

DEMONSTRATION

1. The formation of colloidal systems by the addition of smoke from factories to the atmosphere may be observed from a large number of schools. Specific 35mm slides may be prepared demonstrating the additions from automobiles and the foam found on most waterways below municipal sewage discharge outlets. An attempt should be made to identify the type of particle being emitted from each source. Other slides of smoke, starch suspension, colored gems, clouds, mayonnaise, jellies, foams, and pumice should be examined along with the actual article to illustrate the nine different colloidal systems.

2. Prepare 1,000 mls of 0.1M NaCl, 0.1M $CuSO_4$, 0.1M $NiCl_2$, As_2S_3 sol, $Fe(OH)_3$ sol, Ag sol, and water and place in clear bottles. In a darkened room illuminate each bottle with a strong light source. The solutions will show nothing whereas the colloidal systems will show a dispersion cone. The best light source is a slide projector with a slide having a pinhole in the center to emit light.

3. Other demonstrations may be found in Alyea and Dutton, *Tested Demonstrations in Chemistry*.

STUDENT REFERENCES

Cotton, F. A., and L. D. Lynch, *Chemistry: An Investigative Approach* (Menlo Park, Calif.: Houghton Mifflin Company, 1968), Chapter 10.

Chemical Bond Approach Project, *Chemical Systems* (New York: McGraw-Hill Book Company, Inc., 1964), Chapter 8.

O'Conner, P. R., et al., *Chemistry: Experiments and Principles* (Lexington, Mass.: D. C. Heath & Company, 1968), Chapter 5.

Parry, R. W., et al., *Chemistry: Experimental Foundations* (Englewood Cliffs, N.J.: Prentice-Hall, Inc., 1970), Chapter 5.

TEACHER REFERENCES

Parfitt, G. D., *Surface and Colloid Chemistry* (Elmsford, N.Y.: Pergammon Press, Inc., 1966).

Schmidt, F. C., et al., *General Chemistry* (Lexington, Mass.: D. C. Heath & Company, 1972), Chapter 14.

Vold, M. J., and R. D. Vold, *Colloid Chemistry* (New York: Van Nostrand Reinhold Company, 1964).

"The Teaching of Colloid and Surface Chemistry," *Journal of Chemical Education*, 39:166-195 (1962).

16mm FILMS
Colloids—Encyclopedia Britannica (b/w, sound, 11 min).
The Colloidal State—Coronet Films (color, sound, 16 min).

8mm FILMS
Brownian Motion—International Communications Films (color, silent, 3 min).
Flocculation of Sols—L B Films (color, silent, 4 min).

Acids and Bases

OBJECTIVE
The student shall be able to classify a series of chemical formulas into the following categories: Lewis Acid, Lewis Base, Lowery Acid, Lowery Base, Arrhenius Acid, Arrhenius Base, and others. He must also be able to write the formula for the products of a given acid-base reaction.

DEMONSTRATION
1. The teacher may choose to test a variety of solutions with litmus paper. Among those solutions that are recommended are: fruits of all kinds, vegetables, water from various sources, saliva, and soft drinks. Students should be asked to predict in advance which solutions will be acidic and which will be basic.
2. A small piece of each of the following elements should be placed in the bottom of separate test tubes: Mg, Al, Zn, Cu, and Fe. Two mls of 0.1M HCl should then be added to each test tube. If necessary, the tubes may be placed on the stage of an overhead projector to provide a better view of each reaction. The students should be able to rank each element according to its reactivity with HCl. The process can be repeated with H_2SO_4 to show that this is a typical acid reaction.
3. For other demonstrations see Alyea and Dutton, *Tested Demonstrations in Chemistry*, pp. 11-12 and 61-62.

STUDENT REFERENCES
Cotton, F. A., and L. D. Lynch, *Chemistry: An Investigative Approach* (Boston: Houghton Mifflin Company, 1968), Chapter 16.
Chemical Bond Approach Project, *Chemical Systems* (New York: McGraw-Hill Book Company, Inc., 1964), Chapter 16.
O'Conner, P. R., et al., *Chemistry: Experiments and Principles* (Lexington, Mass.: D. C. Heath & Company, 1968), Chapter 14.

Parry, R. W., et al., *Chemistry: Experimental Foundations* (Englewood Cliffs, N.J.: Prentice-Hall, Inc., 1970), Chapter 13.

TEACHER REFERENCES
Cotton, F. A., and L. D. Lynch, *Teacher's Guide to Chemistry: An Investigative Approach* (Boston: Houghton Mifflin Company, 1968), Chapter 16.
Drago, R. S., and N. A. Matwiyoff, *Acids and Bases* (Lexington, Mass.: D. C. Heath & Company, 1968).
Schmidt, F. C., et al., *General Chemistry* (Lexington, Mass.: D. C. Heath & Company, 1972), Chapter 21.
Vander Werf, C. A., *Acids, Bases and the Chemistry of the Covalent Bond* (New York: Van Nostrand Reinhold Company, 1961).

16mm FILMS
Acid-Base Indicators—Modern Learning Aids (color, sound, 19 min).
Acids, Bases, and Salts—Coronet Films (color, sound, 21 min).

8mm FILMS
Acid-Base Indicators—Modern Learning Aids (color, silent, 4 min).
pH Meter—Harper & Row, Publishers (color, silent, 4 min).
Buffer Solutions—Harper & Row, Publishers (color, silent, 4 min).

Equilibria

OBJECTIVE
The student shall demonstrate an understanding of the equilibrium process by correctly stating how to produce more reactants or products in a given chemical system that includes the chemical equation.

DEMONSTRATION
1. The following procedure will show that both reactants and products exist in solution at the same time.
    a. Prepare solutions of 0.002M KSCN and 0.2M $Fe(NO_3)_3$.
    b. Half fill four large test tubes with the KSCN solution and add three drops of the $Fe(NO_3)_3$ solution.
    c. Leave one test tube as the control and to a second one add a small quantity of KSCN crystals; to a third add three more drops of the $Fe(NO_3)_3$ solution, and to the last test tube add a small quantity of Na_2HPO_4.
    d. Have the students hypothesize what is causing the observed effects in each solution. If possible, they should identify the ions or compounds that are being increased in concentration.

2. The rate of the reaction as a function of concentration can be illustrated by the iodine clock reaction.
   a. Prepare a KlO_3 solution by adding 1 gram of KlO_3 to 500 ml of water and a starch solution by adding 1 gram of starch to 250 ml of boiling water.
   b. Into a 250 ml beaker add 100 ml of each solution.
   c. Into a 250 ml beaker pour solutions made by adding 50 mls of each original solution to 50 additional mls of water.
   d. Into a 250 ml beaker pour solutions made by adding 25 mls of each original solution to 75 mls of additional water.
   e. Determine the time it takes for the color to appear in each mixture. How does the concentration affect the rate of the reaction? (The starch solution should be prepared fresh daily.)

3. The common ion effect on the solubility equilibrium may be seen by adding 10 ml of .1M NaCl to 10 ml of $AgNO_3$ and filtering and adding an additional 10 ml of the .1M NaCl to the filtrate.

4. Additional demonstrations are described in Alyea and Dutton, *Tested Demonstrations in Chemistry*.

STUDENT REFERENCES

Cotton, F. A., and L. D. Lynch, *Chemistry: An Investigative Approach* (Boston: Houghton Mifflin Company, 1968), Chapters 14 and 15.
Chemical Bond Approach Project, *Chemical Systems* (New York: McGraw-Hill Book Company, Inc., 1964), Chapter 15.
O'Conner, P. R., et al., *Chemistry: Experiments and Principles* (Lexington, Mass.: D. C. Heath & Company, 1968), Chapter 13.
Parry, R. W., et al., *Chemistry: Experimental Foundations* (Englewood Cliffs, N.J.: Prentice-Hall, Inc., 1970), Chapters 11, 12, and 13.

TEACHER REFERENCES

Bard, A. J., *Chemical Equilibrium* (New York: Harper & Row Publishers, 1966).
Blackburn, T., *Equilibrium: A Chemistry of Solutions* (New York: Holt, Rinehart and Winston, Inc., 1969).
Cotton, F. A., and L. D. Lynch, *Teacher's Guide to Chemistry: An Investigative Approach* (Boston: Houghton Mifflin Company, 1968), Chapter 14 and 15.
Hamm, R. E., and Carl J. Nyman, *Chemical Equilibrium* (Lexington, Mass.: D. C. Heath & Company, 1968).
Harris, G. M., *Chemical Kinetics* (Lexington, Mass.: D. C. Heath & Company, 1966).
King, E. L., *How Chemical Reactions Occur* (Menlo Park, Calif.: W. A. Benjamin, Inc., 1963).
Morris, Kelso, *Principles of Chemical Equilibrium* (New York: Van Nostrand Reinhold Company, 1967).
Schmidt, F. C., *General Chemistry* (Lexington, Mass.: D. C. Heath & Company, 1972), Chapters 17, 30 and 31.

16mm FILMS
Equilibrium—Modern Learning Aids (color, sound, 24 min).
Solubility Product—Association Instructional Materials (color, sound, 7 min).
Ionization and Ionic Equilibrium—Indiana University (color, sound, 16 min).

8mm FILMS
Liquid/Base Equilibrium—Nuffield Foundation (b/w, silent, 2-5 min).
Solid/Liquid Equilibrium—Nuffield Foundation (b/w, silent, 2-5 min).

Electrochemistry: Oxidation-Reduction Reactions

OBJECTIVES
Given the reactants and the conditions for the reactions, the student shall be able to predict the products and balance the equation.

Given two half-cell reactions, the student shall be able to predict the current flow and the EMF of the cell.

DEMONSTRATION
1. See "Acids and Bases," demonstration number 2.
2. Prepare beakers containing: (a) .1M $ZnSO_4$ and a zinc electrode, (b) 0.1M $CuSO_4$ and a copper electrode, (c) 0.1M $Al(NO_3)_3$ and an aluminum electrode, and (d) 0.1M $Ni(NO_3)_2$ and a nickel electrode. Put a salt bridge between any two beakers in turn and measure the potential difference between the two cells. Compare the data obtained with the predicted values. Prepare a transparency of a typical cell arrangement including the salt bridge and voltmeter before starting the measurements.
3. See also Alyea and Dutton, *Tested Demonstrations in Chemistry*, pp. 13, 20, 87–89, and 124 for additional demonstrations.

STUDENT REFERENCES
Cotton, F. A., and L. D. Lynch, *Chemistry: An Investigative Approach* (Boston: Houghton Mifflin Company, 1968), Chapter 17.

Chemical Bond Approach Project, *Chemical Systems* (New York: McGraw-Hill Book Company, Inc., 1964), Chapters 4, 13, and 14.

O'Conner, P. R., et al., *Chemistry: Experiments and Principles.* Lexington, Mass.: D. C. Heath & Company, 1968), Chapter 15.

Parry, R. W., et al., *Chemistry: Experimental Foundations* (Englewood Cliffs, N.J.: Prentice-Hall, Inc., 1970), Chapter 14.

TEACHER REFERENCES
Cotton, F. A., and L. D. Lynch, *Teacher's Guide to Chemistry: An Investigative Approach* (Boston: Houghton Mifflin Company, 1968), Chapter 17.

Lyons, E. H., Jr., *Introduction to Electrochemistry* (Lexington, Mass.: D. C. Heath & Company, 1967).

Moeller, T., and R. O'Conner, *Ions in Squeous Systems* (New York: McGraw-Hill Book Company, Inc., 1970).
Murray, R. W., and C. N. Reilley, *Electroanalytical Principles* (New York: John Wiley & Sons, Inc., 1963).
Schmidt, F. C., et al., *General Chemistry* (Lexington, Mass.: D. C. Heath & Company, 1972), Chapters 15 and 20.

16mm FILMS
Oxidation-Reduction—Sutherland Educational Films (color, sound, 16 min).
The Development of Electrochemistry—International Film Bureau (color, sound, 19 min).
Faraday's Law—McGraw-Hill Book Company, Inc. (color, sound, 16 min).
Nickel Plating—Royal Institute of Chemistry (color, sound, 27 min).
Electrochemical Cells—Modern Learning Aids (color, sound, 22 min).
The Electromotive Force Series—McGraw-Hill Book Company, Inc. (color, sound, 12 min).

8mm FILMS
An Electrochemical Cell (Animated Mechanism)—Modern Learning Aids (color, silent, 4 min).
A Copper-Silver Electrochemical Cell—Modern Learning Aids (color, silent, 4 min).
Galvanic Cells: Half Cell Reactions—Harper & Row Publishers (color, silent, 4 min).
Lead-Acid Storage Battery—International Communication Films (color, silent, 3 min).
Voltaic Cells—International Communication Films (color, silent, 4 min).

Reaction Mechanisms and Kinetics

OBJECTIVE
The student shall be able to construct the rate equation for a given chemical reaction and detail the effects on the rate of that reaction of concentration changes, temperature, catalysts, and pressure changes.

DEMONSTRATION
1. See *Equilibria* demonstration number 2. An extension of this experiment may be used here. If 50 ml quantities of each solution are heated and reacted at $10°C$ intervals, the temperature effects on the reaction may be noted. Many laboratory manuals provide experiments demonstrating the rate of reaction for various compounds.
2. A molecular model of ethane can be used with Cl_2 to demonstrate what is meant by a mechanistic approach to a reaction.
3. See also Alyea and Dutton, *Tested Demonstrations in Chemistry*, pp. 19, 84–85, and 130 for other demonstrations.

STUDENT REFERENCES

Cotton, F. A., and L. D. Lynch, *Chemistry: An Investigative Approach* (Boston: Houghton Mifflin Company, 1968), Chapter 13.

Chemical Bond Approach Project, *Chemical Systems* (New York: McGraw-Hill Book Company, Inc., 1964), Chapter 8.

O'Conner, P. R., et al., *Chemistry: Experiments and Principles* (Lexington, Mass.: D. C. Heath & Company, 1968), Chapter 12.

Parry, R. W., et al., *Chemistry: Experimental Foundations* (Englewood Cliffs, N.J.: Prentice-Hall, Inc., 1970), Chapter 10.

TEACHER REFERENCES

Abbott, D., *An Introduction to Reaction Kinetics* (Boston: Houghton Mifflin Company, 1968).

Cotton, F. A., and L. D. Lynch, *Teacher's Guide to Chemistry: An Investigative Approach* (Boston: Houghton Mifflin Company, 1968), Chapter 13.

Edwards, J. O., *Inorganic Reaction Mechanisms: An Introduction* (Menlo Park, Calif.: W. A. Benjamin, Inc., 1969).

Harris, G. M., *Chemical Kinetics* (Lexington, Mass.: D. C. Heath & Company, 1968).

Menger, R. M., *Problems in Organic Reaction Mechanism* (New York: Appleton-Century-Crofts, 1968).

Schmidt, F. C., *General Chemistry* (Lexington, Mass.: D. C. Heath & Company, 1972), Chapters 8 and 17.

Sykes, A. G., *Kinetics of Inorganic Reactions* (Elmsford, N.Y.: Pergamon Press, Inc., 1966).

16mm FILMS

Speed of Chemical Change—Film Associated of California (color, sound, 15 min).
Catalysis—Modern Learning Aids (color, sound, 17 min).
Mechanism of an Organic Reaction—Modern Learning Aids (color, sound, 20 min).

8mm FILMS

Reaction Kinetics—Harper & Row, Publishers (color, silent, 4 min).
Inversion, Retention and Racemization—Harper & Row, Publishers (color, silent, 4½ min).

Thermodynamics

OBJECTIVE

Given a balanced chemical reaction and a table of standard molar enthalpies of formation, the student shall be able to calculate the enthalpy change of a reaction and to state whether the reaction will be endothermic or exothermic with respect to this reaction.

DEMONSTRATION
1. Exothermic reactions may be demonstrated by the burning of paper or magnesium ribbon, mixing solutions of HCl and NaOH, and noting the temperature rise. In addition, wire coiled for about 3 inches at its end can be attached to one terminal of a door bell, and a piece of copper wire attached to a piece of magnesium ribbon 3 inches long can be attached to the other bell terminal. Immersion of the coil of copper and magnesium into a beaker of dilute sulfuric acid will cause the bell to ring.
2. Endothermic reactions may be demonstrated by the addition of a small quantity of water to methyl chloride in a beaker. The water freezes almost instantly and a dramatic effect can be shown by quickly turning the beaker upside down. Fill a 50 ml beaker with ammonium nitrate and place it on a wet spot on the demonstration table. Add 50 ml of water and stir. The beaker will freeze to the desk.
3. A variety of other demonstrations may be found in Alyea and Dutton, *Tested Demonstrations in Chemistry*.

STUDENT REFERENCES
Cotton, F. A., and L. D. Lynch, *Chemistry: An Investigative Approach* (Boston: Houghton Mifflin Company, 1968), Chapter 12.
Chemical Bond Approach Project, *Chemical Systems* (New York: McGraw-Hill Book Company, Inc., 1964), Chapter 15.
O'Conner, P. R., et al., *Chemistry: Experiments and Principles* (Lexington, Mass.: D. C. Heath & Company, 1968), Chapter 11.
Parry, R. W., et al., *Chemistry: Experimental Foundations* (Englewood Cliffs, N.J.: Prentice-Hall, Inc., 1970), Chapter 9.

TEACHER REFERENCES
Augrist, S. W., "Perpetual Motion Machines," *Scientific American* (Jan., 1968), p. 114.
Bearman, R. J., and Benjamin Chu, *Problems in Chemical Thermodynamics* (Reading, Mass.: Addison-Wesley Publishing Co., Inc., 1967).
Cotton, F. A., and L. D. Lynch, *Teacher's Guide to Chemistry: An Investigative Approach* (Boston: Houghton Mifflin Company, 1968), Chapter 12.
Franzen, H. F., and B. C. Gerstein, *Rudimentary Chemical Thermodynamics* (Lexington, Mass.: D. C. Heath & Company, 1971).
Mahan, B. H., *Elementary Chemical Thermodynamics* (Menlo Park, Calif.: W. A. Benjamin, Inc., 1963).
Matthews, G. W. J., "Demonstrations of Spontaneous Endothermic Reactions," *Journal of Chemical Education*, 43:476 (1966).
Nash, L. K., *Elements of Chemical Thermodynamics* (Reading, Mass.: Addison-Wesley Publishing Co., Inc., 1962).
Nash, L. K., "Chemical Equilibrium as a State of Maximal Entropy," *Journal of Chemical Education*, 47:353 (1970).

Pimental, G. C., and R. D. Spratley, *Understanding Chemical Thermodynamics* (San Francisco: Holden-Day, Inc., 1969).
Schmidt, F. C., et al., *General Chemistry* (Lexington, Mass.: D. C. Heath & Company, 1972), Chapter 18.
Waser, W. A., *Basic Chemical Thermodynamics* (Menlo Park, Calif.: W. A. Benjamin Co., 1966).

16mm FILMS
Chaos and Evolution—Peter M. Roebeck and Co. (b/w, sound, 30 min).
Equilibrium—The Limit of Disorder—Peter M. Roebeck and Co. (b/w, sound, 30 min).
Molecules in Motion—Peter M. Roebeck and Co. (b/w, sound, 30 min).
The Second Law—Peter M. Roebeck and Co. (b/w, sound, 30 min).
Entropy—Peter M. Roebeck and Co. (b/w, sound, 30 min).

8mm FILMS
Energy Conversion—Harper & Row, Publishers (color, silent, 4 min).
Heat and Fusion—Harper & Row, Publishers (color, silent, 4 min).
Energy Cycles—Willard Grant Press (color, silent, 4½ min).
Energy Transformation I and II—UNESCO (color, silent, 3 min and 2 min).
Physical and Chemical Changes: Very Fast Reaction—Encyclopedia Britannica (color, silent, 3½ min).

Periodic Properties of the Elements

OBJECTIVE
The student shall be able to list the criteria on which attempts have been made to classify the elements into groups. He shall also be able to state the chemical properties of any given family within the table and to discuss the trends of the physical properties of the elements that are dependent upon size, nuclear charge, and the like.

DEMONSTRATION
1. The teacher can prepare a number of styrofoam spheres with toothpicks sticking out varying from one to eight to represent the number of electrons in the outermost energy levels. Different sized spheres can be used and a series of spheres of different colors for each period can also be used. The objective is to have the students classify these spheres according to the number of toothpicks sticking out. The first criteria can be classification that would correspond to a period. The spheres can then be grouped according to the number of toothpicks that corresponds to the number of electrons in the outermost energy level.

2. Family properties can be demonstrated by examining the formulas of the elements of one family with the same common anion. If families one, two, and three are presented in combination with the chloride ion the students should be able to correlate the number of electrons in the outermost energy level.

3. Demonstrations that may be used to illustrate the properties of individual elements may be found in Alyea and Dutton, *Tested Demonstrations in Chemistry*.

STUDENT REFERENCES

Cotton, F. A., and L. D. Lynch, *Chemistry: An Investigative Approach* (Boston: Houghton Mifflin Company, 1968), Chapter 6.

Chemical Bond Approach Project, *Chemical Systems* (New York: McGraw-Hill Book Company, Inc., 1964), Chapter 10.

O'Conner, P. R., et al., *Chemistry: Experiments and Principles* (Lexington, Mass.: D. C. Heath and Co., 1968), Chapter 7.

Parry, R. W., et al., *Chemistry: Experimental Foundations* (Englewood Cliffs, N.J.: Prentice-Hall, Inc., 1970), Chapter 8.

TEACHER REFERENCES

Cotton, F. A., and L. D. Lynch, *Teacher's Guide to Chemistry: An Investigative Approach* (Boston: Houghton Mifflin Company, 1968), Chapter 6.

Mechamkin, Howard, *The Chemistry of the Elements* (New York: McGraw-Hill Book Company, Inc., 1968).

Rich, R. L., *Periodic Correlations—Physical Inorganic Chemistry Series* (Menlo Park, Calif.: W. A. Benjamin, Inc., 1965).

Schmidt, F. C., et al., *General Chemistry* (Lexington, Mass.: D. C. Heath & Company, 1972), Chapter 3.

Seaborg, Glenn, *Man-Made Transuranium Elements* (Englewood Cliffs, N.J.: Prentice-Hall, Inc., 1963).

Sisler, H. H., *Electronic Structure, Properties and the Periodic Law* (New York: Van Nostrand Reinhold Company, 1963).

16mm FILMS

Chemical Families—Modern Learning Aids (color, sound, 22 min).
Electronegativity—Association Instructional Materials (color, sound, 4 min).
Melting Points—Determination and Trends—Association Instructional Materials (color, sound, 9 min).
Metals and Non-Metals—Association Instructional Materials (color, sound, 9 min).

8mm FILMS

Alkali Metal Reactions with Chlorine and with Water—Modern Learning Aids (color, silent, 4 min).
Group 1A Elements: Physical Properties, Part 1—Harper & Row, Publishers (color, silent, 3½ min).

Group 1A Elements: Physical Properties, Part 2—Harper & Row, Publishers (color, silent, 3½ min).
Group 1A Elements: Physical and Chemical Properties—Harper & Row, Publishers (color, silent, 4 min).

Elements: Family IA and IIA

OBJECTIVE

The student shall be able to list the elements of each family in order of reactivity, size, and electronegativity as well as write the formulas for the oxides and the chlorides of each element.

DEMONSTRATION

1. The order of reactivity can be established for Family IA by adding small pieces of Li, Na, and K to separate beakers of water. Caution: Use only small quantities of the metals as the hydrogen evolved is flammable and may explode.
2. Prepare transparencies comparing the relative size of each element to the other members of the same group.
3. Many demonstrations may be found for each metal by looking in *Tested Demonstrations in Chemistry*, by Alyea and Dutton, under the element.

STUDENT REFERENCES

Cotton, F. A., and L. D. Lynch, *Chemistry: An Investigative Approach* (Boston: Houghton Mifflin Company, 1968), Chapter 19.

O'Conner, P. R., et al., *Chemistry: Experiments and Principles* (Lexington, Mass.: D. C. Heath & Company, 1968), Chapter 7.

Parry, R. W., et al., *Chemistry: Experimental Foundations* (Englewood Cliffs, N.J.: Prentice-Hall, Inc., 1970), Chapter 8 and 20.

TEACHER REFERENCES

Cotton, F. A., and L. D. Lynch, *Teacher's Guide to Chemistry: An Investigative Approach* (New York: Houghton Mifflin Company, 1968), Chapter 19.

Latimer, W., and J. Hildebrand, *Reference Book of Inorganic Chemistry* (New York: The Macmillan Company, 1965).

Mott, N. F., and H. Jones, *The Theory of the Properties of Metals and Alloys* (New York: Dover Publications, Inc., 1968).

Schmidt, F. C., et al., *General Chemistry* (Lexington, Mass.: D. C. Heath & Company, 1972), Chapters 16, 34, and 35.

Steele, David, *The Chemistry of the Metallic Elements* (Elmsford: Pergamon Press, Inc., 1966).

16mm FILMS
Chemical Families—Modern Learning Aids (color, sound, 22 min).
The Sodium Family—Coronet Films (color, sound, 16 min).

8mm FILMS
Alkali Metal Reactions with Chlorine and with Water—Modern Learning Aids (color, silent, 4 min).
Group 1A Elements: Physical Properties, Part 1 and 2—Harper & Row, Publishers (color, silent, 3½ min).
Group 1A Elements: Chemical Properties—Harper & Row, Publishers (color, silent, 3½ min).

Elements: Family IIIA and IVA

OBJECTIVE
The student shall be able to list the elements of each family in order of increasing atomic weight, atomic radius, and reactivity with the halogens.

DEMONSTRATIONS
1. Drop crystals of $CO(NO_3)_2$, $Pb(NO_3)_2$, $Ni(NO_3)_2$, $Ca(NO_3)_2$, $Cu(NO_3)_2$ and $Mn(NO_3)_2$ into a 1,000 ml beaker containing 150 ml of sodium silicate and 600 ml of water. Be sure to spread a thin layer of sand in the bottom of the beaker before adding the crystals. Brightly colored silicates will form in the liquid within a short time.
2. See Alyea and Dutton, *Tested Demonstrations in Chemistry*, for other demonstrations that are listed under the name of the individual element.
3. A transparency can be prepared illustrating the crystal structure of boron and carbon to show one of their physical properties.

STUDENT REFERENCES
Cotton, F. A., and L. D. Lynch, *Chemistry: An Investigative Approach* (Boston: Houghton Mifflin Company, 1968), Chapter 19.
O'Conner, P. R., et al., *Chemistry: Experiments and Principles* (Lexington, Mass.: D. C. Heath & Company, 1968), Chapter 7.
Parry, R. W., et al., *Chemistry: Experimental Foundations* (Englewood Cliffs, N.J.: Prentice-Hall, Inc., 1970), Chapter 8.

TEACHER REFERENCES
Addison, W. E., *Structural Principles in Organic Compounds* (New York: John Wiley & Sons, Inc., 1961).
Cotton, F. A., and L. D. Lynch, *Teacher's Guide to Chemistry: An Investigative Approach* (Boston: Houghton Mifflin Company, 1968), Chapter 19.

Jolly, W. L., *The Chemistry of the Non-Metals* (Englewood Cliffs, N.J.: Prentice-Hall, Inc., 1966).

Rochow, E. G., *The Metalloids* (Lexington, Mass.: D. C. Heath & Company, 1966).

Schmidt, F. C., et al., *General Chemistry* (Lexington, Mass.: D. C. Heath & Company, 1972), Chapters 26, 28, 38.

16mm FILMS

The Modern Chemist—Diamond Synthesis—Sutherland Educational Films (color, sound, 13 min).

Silicon and Its Compounds—Coronet Films (color, sound, 14 min).

Aluminum—Metal of Many Faces—(color, sound, 28 min).

8mm FILMS

Corrosion III—Aluminum—Willard Grant Press (color, silent, 4 min).

Elements: Family VA and VIA

OBJECTIVE

The student shall be able to list the elements of these families in order of increasing atomic weight and be able to list the sources of such elements as oxygen and sulfur.

DEMONSTRATIONS

1. The properties of oxygen as well as a typical decomposition reaction to produce oxygen can be demonstrated by heating a mixture of $KClO_3$ and MnO_2 in a large test tube. The procedure for collecting oxygen is demonstrated in most laboratory manuals. The density of oxygen with respect to air can be established with a glowing splint and two bottles of collected oxygen. The splint will burst into flame in the bottle filled with oxygen.

2. The solubility of ammonia in water can be shown with a demonstration as indicated in Figure 2-2. The flask should be completely filled with ammonia gas and closed with a two-hole stopper with the appropriate glassware inserted. Stick the other end of the tubing almost to the bottom of a 2,000 ml beaker that is three-quarters full of water. Place strips to adhesive tape around portions of the flask to guard against flying glass when the stopcock is open. Open the stopcock and watch the water spew from the end of the tube in the flask. It may be necessary to squirt a little of the mixture of water and phenolphthalein into the flask to start the fountain flowing. The color change should be a source for several questions.

3. Other demonstrations are listed under each element in Alyea and Dutton, *Tested Demonstrations in Chemistry*.

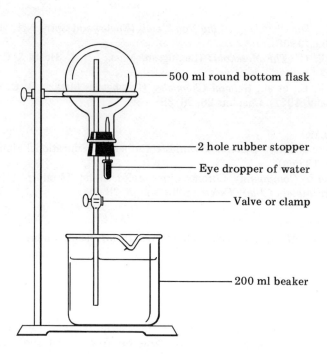

Figure 2-2

STUDENT REFERENCES
Cotton, F. A., and L. D. Lynch, *Chemistry: An Investigative Approach* (Boston: Houghton Mifflin Company, 1968), Chapter 19.
O'Conner, P. R., et al., *Chemistry: Experiments and Principles* (Lexington, Mass.: D. C. Heath & Company, 1968), Chapter 7.
Parry, R. W., et al., *Chemistry: Experimental Foundations* (Englewood Cliffs, N.J.: Prentice-Hall, Inc., 1970), Chapter 8.

TEACHER REFERENCES
Johnson, R. C., *Introductory Descriptive Chemistry: Selected Non-Metals, Their Properties and Behavior* (Menlo Park, Calif.: W. A. Benjamin, Inc., 1967).
Jolly, W. L., *The Inorganic Chemistry of Nitrogen* (Menlo Park, Calif.: W. A. Benjamin, Inc., 1966).
Schmidt, F. C., et al., *General Chemistry* (Lexington, Mass.: D. C. Heath & Company, 1972), Chapters 6, 22, 24, 25, and 26.
Sherwin, E., and G. J. Weston, *The Chemistry of the Non-Metallic Elements* (Elmsford, N.Y.: Pergamon Press, Inc., 1966).

16mm FILMS
Nitric Acid—Modern Learning Aids (color, sound, 18 min).

Nitrogen and Ammonia—Coronet Films (color, sound, 16 min).
Phosphorus—Coronet Films (color, sound, 19 min).
Oxygen—Coronet Films (color, sound, 11 min).
Sulfur and Its Compounds—Coronet Films (color, sound, 14 min).
Preparation of Oxygen: 7 Demonstrations—Encyclopedia Britannica (color, sound, 15 min).

8mm FILMS
Combustion II: Burning Phosphorus—McGraw-Hill (color, silent, 3 min).
Ammonia Fountain—Encyclopedia Britannica (color, silent, 4 min).
Preparation of Oxygen (series of 7)—Encyclopedia Britannica (color, silent, 4 min).

Elements: Family VII and VIII

OBJECTIVE
The student shall be able to draw Lewis structures of the polyatomic ions of any member of the halogen family and oxygen, and rank them in order of their reactivity in replacement reactions.

DEMONSTRATION
1. The relative reactive powers of the halogens can be seen in the replacement reactions where 5 ml of chlorine water is added to a 0.1M solution of NaBr in a 50 ml beaker sitting on the stage of an overhead projector. This may be repeated by adding 5 ml of bromine water to a 0.1M NaI solution. A few crystals of I_2 can be added to the solutions of NaCl, NaBr, and NaI to show the lack of reaction.
2. A transparency showing the XeF_4 square planar should be used when discussing the surprising reactive capability of the inert gas, xenon.
3. Demonstrations for the generation of the halogen gases are available in Alyea and Dutton, *Tested Demonstrations in Chemistry*, pp. 102-103.

STUDENT REFERENCES
Cotton, F. A., and L. D. Lynch, *Chemistry: An Investigative Approach* (Boston: Houghton Mifflin Company, 1968), Chapter 18.
O'Conner, P. R., et al., *Chemistry: Experiments and Principles* (Lexington, Mass.: D. C. Heath & Company, 1968), Chapter 19.
Parry, R. W., et al., *Chemistry: Experimental Foundations* (Englewood Cliffs, N.J.: Prentice-Hall, Inc., 1970), Chapter 20.

TEACHER REFERENCES
Clausen, H. H., *The Noble Gases* (Lexington, Mass.: D. C. Heath & Company, 1966).

Cotton, F. A., and L. D. Lynch, *Teacher's Guide to Chemistry: An Investigative Approach* (Boston: Houghton Mifflin Company, 1968), Chapter 18.
Frey, John E., "Discovery of the Noble Gases and Foundations of the Theory of Atomic Structure," *Journal of Chemical Education*, 42, 371-374 (1966).
Jolly, W. L., *The Chemistry of the Non-Metals* (Englewood Cliffs, N.J.: Prentice-Hall, Inc., 1966).
Schmidt, F. C., et al., *General Chemistry* (Lexington, Mass.: D. C. Heath & Company, 1972), Chapters 19 and 23.

16mm FILMS
The Family of Halogens—McGraw-Hill Book Company, Inc. (color, sound, 13 min).
Chlorine—A Representative Halogen—Sutherland Educational Films (color, sound, 15 min).
Bromine—Element from the Sea—Modern Learnings Aids (color, sound, 22 min).
A Research Problem: Inert (?) Gas Compounds—Modern Learning Aids (color, sound, 19 min).
Halogens—Coronet Films (color, sound, 14 min).
Inert Gases—Royal Institute of Chemistry (color, sound, 11 min).

Elements: Transition

OBJECTIVE
The student shall be able to list the elements in each row of the transition series in order of increasing weights. The student shall also be able to list the most common oxidation states of each element and which oxidation state is most likely to occur.

DEMONSTRATION
1. The colors of the different oxidation states of the same element may be demonstrated by the following compounds that should be available for the student to observe: $CrSO_4$, blue; $Cr_2(SO_4)_3$, purple; K_2CrO_4, yellow; and $K_2Cr_2O_7$, orange. Also use $MnSO_4$, pink; MnO_2, black; $KMnO_4$, purple.
2. Transparencies should be prepared indicating the approximate ionic and atomic radii for each element in each series.
3. Depending upon the teacher's own objectives, demonstrations from Alyea and Dutton, *Tested Demonstrations in Chemistry*, pp. 104-105 may be used.

STUDENT REFERENCES
Cotton, F. A., and L. D. Lynch, *Chemistry: An Investigative Approach* (Boston: Houghton Mifflin Company, 1968), Chapters 21 and 22.

O'Conner, P. R., et al., *Chemistry: Experiments and Principles* (Lexington, Mass.: D. C. Heath & Company, 1968), Chapter 20.

Parry, R. W., et al., *Chemistry: Experimental Foundations* (Englewood Cliffs, N.J.: Prentice-Hall, Inc., 1970), Chapter 21.

TEACHER REFERENCES

Cotton, F. A., and L. D. Lynch, *Teacher's Guide to Chemistry: An Investigative Approach* (Boston: Houghton Mifflin Company, 1968), Chapters 21 and 22.

Dorain, P. B., *Symmetry in Inorganic Chemistry* (Reading, Mass.: Addison-Wesley Publishing Co., Inc., 1965).

Larsen, E. M., *Transitional Elements* (Menlo Park, Calif.: W. A. Benjamin, Inc., 1965).

Murmann, R. K., *Inorganic Complex Compounds* (New York: Van Nostrand Reinhold Company, 1964).

Schmidt, F. C., et al., *General Chemistry* (Lexington, Mass.: D. C. Heath & Company, 1972), Chapters 36, 37, 41, and 42.

16mm FILMS

Vanadium—A Transition Element—Modern Learning Aids (color, sound, 22 min).
Chromium and Manganese—Coronet Films (color, sound, 38 min).

Nuclear Chemistry

OBJECTIVE

The student shall be able to define radioactivity in terms of the three different types of radiation. The student shall also demonstrate an understanding of nuclear energy potential by being able to list arguments for and against the use of nuclear power plants to generate electricity. These arguments shall demonstrate an understanding of the concepts of chain reactions, fission, fusion, critical mass, control rods, and the like.

DEMONSTRATION

1. Obtain a commercial cloud chamber and follow the directions to demonstrate the paths of gamma radiation and cosmic radiation.

2. A wrapped piece of photographic film can be exposed to a piece of uranium ore and developed. This will demonstrate the penetrating power of the radiation.

3. Use uranyl nitrate as a beta source in conjunction with a Geiger counter, counter rate meter, pieces of paper, sheets of aluminum foil, sheets of copper foil, and sheets of lead foil. The quantity of source material should be adjusted to provide approximately 1,000 counts per minute. The distance between the source and the probe should be approximately five inches. Insertion of varying quantities

of the shielding material will demonstrate the penetrating power of beta radiation. Discussion should lead to an examination of the radiation shielding that is needed at nuclear power plants.

4. Alyea and Dutton, *Tested Demonstrations in Chemistry*, pp. 87, 89, 104, 120, 131, 185, lists other demonstrations the teacher may wish to use.

STUDENT REFERENCES
Cotton, F. A., and L. D. Lynch, *Chemistry: An Investigative Approach* (Boston: Houghton Mifflin Company, 1968), Chapter 22.
Chen, F. F., "The Leakage Problem in Fusion Reactors," *Scientific American*, July, 1967; p. 76.
Gough, W. C., and B. J. Eastluna, "The Prospects of Fusion Power," *Scientific American*, February, 1971; p. 50.
O'Conner, P. R., et al., *Chemistry: Experiments and Principles* (Lexington, Mass.: D. C. Heath and Co., 1968), Chapter 21.
Parry, R. W., et al., *Chemistry: Experimental Foundations* (Englewood Cliffs, N.J.: Prentice-Hall, Inc., 1970), Chapter 9.
Seaborg, G. T., and J. L. Bloom, "The Synthetic Elements," *Scientific American*, April 1969, p. 56.
────── and ──────, "Fast Breeder Reactors," *Scientific American*, November, 1970; p. 13.
Swartout, K. A., "Critical Chemical Problems in the Development of Nuclear Reactors," *Journal of Chemical Education*, 45:304 (1968).

TEACHER REFERENCES
Cohen, Bernard L., *The Heart of the Atom* (Garden City, N.Y.: Doubleday & Company, Inc., 1967).
Choppin, Gregory R., *Nuclei and Radioactivity: Elements of Nuclear Chemistry* (Menlo Park, Calif.: W. A. Benjamin, Inc., 1964).
Schmidt, F. C., et al., *General Chemistry* (Lexington, Mass.: D. C. Heath & Company, 1972), Chapter 29.

16mm FILMS
Exploring the Atomic Nucleus—Coronet Films (color, sound, 13½ min).
Atomic Accelerators—Encyclopedia Britannica (b/w, sound, 30 min).
Nuclear Radiation: Uses in Industry—Central Scientific Company (color, sound, 15 min).
The Nuclear Reactor—McGraw-Hill Book Company, Inc. (b/w, sound, 11 min).

8mm FILMS
Radioactive Decay—Harper & Row, Publishers (color, silent, 4 min).
Radioactivity—Harper & Row, Publishers (color, silent, 4 min).
Chain Reaction-Controlled Chain Reaction—Encyclopedia Britannica Films (color, silent, 2½ min).

Coordination Chemistry

OBJECTIVE

The student shall be able to predict the structure of a coordination from the formula of the compound and the ability to determine the number of lone-pair and bond-pair electrons.

DEMONSTRATION

1. Prepare a transparency showing the results of the titration of the following with $AgNO_3$: $Co(NH_3)_6Cl_3$, $Co(NH_3)_5Cl_3$, $Co(NH_3)_4Cl_3$, and $Co(NH_3)_3Cl_3$. The conclusions should lead to two types of bonding taking place in these molecules.

2. Construct molecular models for the different types of structures that are common to coordination compounds. These are: linear, trigonal planar, square planar, tetrahedral, trigonal bipyramid, tetrahedral bipyramid, and so on.

3. Prepare a transparency with the following orbitals for determining the type of hybridization and the number of lone-pair and bond-pair electrons for a given compound (Figure 2-3).

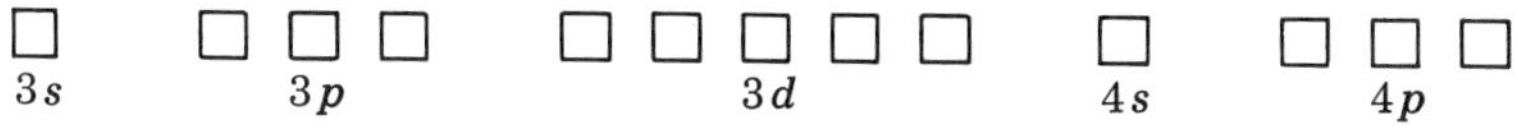

☐   ☐ ☐ ☐   ☐ ☐ ☐ ☐ ☐   ☐   ☐ ☐ ☐
$3s$ $3p$ $3d$ $4s$ $4p$

Figure 2-3

STUDENT REFERENCES

Cotton, F. A., and L. D. Lynch, *Chemistry: An Investigative Approach* (Boston: Houghton Mifflin Company, 1968), Chapter 21.

O'Conner, P. R., et al., *Chemistry: Experiments and Principles* (Lexington, Mass.: D. C. Heath & Company, 1968), Chapter 20.

Parry, R. W., et al., *Chemistry: Experimental Foundations* (Englewood Cliffs, N.J.: Prentice-Hall, Inc., 1970), Chapter 21.

TEACHER REFERENCES

Barrow, G. M., *The Structure of Molecules* (Menlo Park, Calif.: W. A. Benjamin, Inc., 1963).

Basolo, F., and R. C. Johnson, *Coordination Chemistry* (Menlo Park, Calif.: W. A. Benjamin, Inc., 1964).

Cotton, F. A., and L. D. Lynch, *Teacher's Guide to Chemistry: An Investigative Approach* (Boston: Houghton Mifflin Company, 1968), Chapter 21.

Martin, D. F., and B. A. Martin, *Coordination Compounds* (New York: McGraw-Hill Book Company, Inc., 1964).

Murmann, R. K., *Inorganic Complex Compounds* (New York: Van Nostrand Reinhold Company, 1964).

Quagliano, J. V., and L. M. Vallarino, *Coordination Chemistry* (Lexington, Mass.: D. C. Heath & Company, 1968).

Schmidt, F. C., et al., *General Chemistry* (Lexington, Mass.: D. C. Heath & Company, 1972), Chapter 33.

FILMS
See the section Elements: Transition.

Organic Chemistry

OBJECTIVE
The student shall be able to take a group of structural formulas for various organic compounds and correctly classify them according to functional group: alcohol, acid, aldehyde, ketone, ether, ester, as well as aromatic and alliphatic.

DEMONSTRATION
1. Styrofoam or wooden spheres should be constructed as models for the typical alliphatic hydrocarbons. During discussion, the functional groups can be added in place of the hydrogen so the students can visualize the differences. Individual student model kits can also be purchased and it would be useful to have the students construct molecules to fit the various categories.
2. Addition reactions can be demonstrated by adding drops of a solution of dilute bromine in carbon tetrachloride to beakers of cyclohexane, cyclohexene, and benzene. The students should count the number of drops of the bromine solution added to obtain a reaction. The solution should be stirred well after each addition of the bromine solution. For better class visibility, the beakers may be placed on the stage of an overhead projector.
3. Transparencies of the sigma bond, pi bond, and the bond in the benzene ring should be prepared to complement the models of these bonds that are available. The students should be able to explain the ease of addition to an unsaturated compound as compared to the ease of adding to a saturated compound.
4. Alyea and Dutton, *Tested Demonstrations in Chemistry*, pp. 47-48, contains descriptions of other demonstrations that may assist the teacher to help the students conceptualize various concepts.

STUDENT REFERENCES
Cotton, F. A., and L. D. Lynch, *Chemistry: An Investigative Approach* (Boston: Houghton Mifflin Company, 1968), Chapter 23.

Chemical Bond Approach Project, *Chemical Systems* (New York: McGraw-Hill Book Company, Inc., 1964), Chapter 7.

O'Conner, P. R., et al., *Chemistry: Experiments and Principles* (Lexington, Mass.: D. C. Heath & Company, 1968), Chapter 18.

Parry, R. W., et al., *Chemistry: Experimental Foundations* (Englewood Cliffs, N.J.: Prentice-Hall, Inc., 1970), Chapter 18.

TEACHER REFERENCES

Benfrey, O. T., *The Names and Structures of Organic Compounds* (New York: John Wiley & Sons, Inc., 1966).

Campaigne, E., *Elementary Organic Chemistry* (Englewood Cliffs, N.J.: Prentice-Hall, Inc., 1962).

Cotton, F. A., and L. D. Lynch, *Teacher's Guide to Chemistry: An Investigative Approach* (Boston: Houghton Mifflin Co., 1968), Chapter 23.

Ireland, R., *Organic Synthesis* (Englewood Cliffs, N.J.: Prentice-Hall, Inc., 1969).

Schenk, G. H., *Organic Functional Group Analysis: Theory and Development* (Elmsford, N.Y.: Pergamon Press, Inc., 1968).

Schmidt, F. C., et al., *General Chemistry* (Lexington, Mass.: D. C. Heath & Company, 1972), Chapter 26.

Stille, J. K., *Industrial Organic Chemistry* (Englewood Cliffs, N.J.: Prentice-Hall, Inc., 1968).

Traynham, J., *Organic Nomenclature: A Programmed Introduction* (Englewood Cliffs, N.J.: Prentice-Hall, Inc., 1966).

Whitfield, R. C., *A Guide to Understanding Basic Organic Reactions* (Boston: Houghton Mifflin Company, 1969).

16mm FILMS

Carbon and Its Compounds—Coronet Films (color, sound, 11 min).
Hydrocarbons and Their Structures—Coronet Films (color, sound, 13 min).
Mechanism of an Organic Reaction—Modern Learning Aids (color, sound, 20 min).
Many others are available from manufacturers using specific compounds such as petroleum and synthetic fabrics.

8mm FILMS

Reaction Kinetics—Harper & Row, Publishers (color, silent, 4 min).
Inversion, Retention, and Racemization—Harper & Row, Publishers (color, silent, 4 min).

Biochemistry

OBJECTIVE

The student shall be able to trace the path of the three basic groups of compounds that are needed by the body as they pass from their source to the cell. The description of this process should include a general procedure for the breakdown conditions for each type of molecule, transportation mechanism, and the general conditions for the reformation of the compounds in the cell.

DEMONSTRATION
1. Place a 250 ml beaker on an asbestos square on the desk. Cover the bottom of the beaker with about one half inch of sucrose. Add a sufficient amount of concentrated H_2SO_4 to cover the sugar and than wait. The sugar will dehydrate and a cone of carbon will rise in the beaker to a height that depends on the initial quantity of sucrose. Caution: Adequate ventilation should be available as the SO_2 fumes burn the nose.
2. The action of enzymes can be demonstrated by adding the catalase in a few drops of blood to a small quantity of 3 percent H_2O_2, to produce a foaming reaction.
3. Three-dimensional models of proteins, carbohydrates, and fats are available to demonstrate the complexity of these molecules. It is easy to use these models to demonstrate the points within the molecule at which reactions may occur.
4. For further demonstrations see Alyea and Dutton, *Tested Demonstrations in Chemistry*, pp. 47-48.

STUDENT REFERENCES
Cotton, F. A., and L. D. Lynch, *Chemistry: An Investigative Approach* (Boston: Houghton Mifflin Company, 1968), Chapter 24.
Chemical Bond Approach Project, *Chemical Systems* (New York: McGraw-Hill Book Company, Inc., 1964).
O'Conner, P. R., et al., *Chemistry: Experiments and Principles* (Lexington, Mass.: D. C. Heath & Company, 1968), Chapter 22.
Parry, R. W., et al., *Chemistry: Experimental Foundations* (Englewood Cliffs, N.J.: Prentice-Hall, Inc., 1970), Chapter 22.

TEACHER REFERENCES
Barry, M., and E. M. Barry, *An Introduction to the Structure of Biological Molecules* (Englewood Cliffs, N.J.: Prentice-Hall, Inc.), 1969.
Bennett, T. P., and E. Frieden, *Modern Topics in Biochemistry* (New York: The Macmillan Company, 1966).
Cotton, F. A., and L. D. Lynch, *Teacher's Guide to Chemistry: An Investigative Approach* (Boston: Houghton Mifflin Company, 1968), Chapter 24.
DeBey, H. J., *Introduction to Chemistry of Life: Biochemistry* (Reading, Mass.: Addison-Wesley Publishing Co., Inc., 1969).
Routh, J. I., *Introduction to Biochemistry* (Philadelphia: W. B. Saunders Company, 1971).
Schmidt, F. C., et al., *General Chemistry* (Lexington, Mass.: D. C. Heath & Company, 1972), Chapter 27.

16mm FILMS
Biochemistry and Molecular Structure—Modern Learning Aids (color, sound, 22 min).
Chemical Machinery—McGraw-Hill Book Company, Inc. (color, sound, 28 min).
A Cell's Chemical Organization—McGraw-Hill Book Company, Inc. (color, sound, 28 min).
Chemistry of the Cell—I: The Structure of Proteins and Nucleic Acids—McGraw-Hill Book Company, Inc. (color, sound, 21 min).

8mm FILMS
Giant Molecules—Proteins—Royal Institute of Chemistry (color, silent, 5 min).

Environmental Chemistry

OBJECTIVE
The student shall be able to identify specific problems in the overall pollution situation that can be dealt with by using the principles developed in the course.

DEMONSTRATIONS
1. Begin with a beaker of turbulent river water collected downstream from any town. What is contained in the sample and some of the procedures that must be used to clean it up can be demonstrated by filtering the sample, adding silver nitrate solution to the filtrate to show the presence of the chloride ion, and culturing (with care) a sample to show the microscopic life that still exists after the first two treatments.
2. Divide the class into several groups and provide each group with a different sample, such as a bag full of garbage or sludge, a bottle of unknown suspicious-looking liquid, or a bottle of a foul-smelling gas. The objective is for each group to develop a way of disposing of its particular pollution source.
3. A collection of slides can be assembled depicting a large number of pollution situations that are caused by chemical processes and could be corrected by the applications of other chemical principles.

STUDENT REFERENCES
Cotton, F. A., and L. D. Lynch, *Chemistry: An Investigative Approach* (Boston: Houghton Mifflin Company, 1968), Chapter 25.
Parry, R. W., et al., *Chemistry: Experimental Foundations* (Englewood Cliffs, N.J.: Prentice-Hall, Inc., 1970), Chapter 23.

TEACHER REFERENCES
American Chemical Society, *Cleaning Our Environment: The Chemical for Action* (Washington, D.C.: American Chemical Society, 1969).
Cotton, F. A., and L. D. Lynch, *Teacher's Guide to Chemistry: An Investigative Approach* (Boston: Houghton Mifflin Company, 1968), Chapter 25.
Fisher, R. B., *Science, Man and Society* (Philadelphia: W. B. Saunders Company, 1971).
Giddings, J. C., and M. B. Monroe, *Our Chemical Environment* (San Francisco: Canfield Press, 1972).
Goldman, M. I., *Controlling Pollution: The Economics of a Cleaner America* (Englewood Cliffs, N.J.: Prentice-Hall, Inc., 1967).
Schmidt, F. C., et al., *General Chemistry* (Lexington, Mass.: D. C. Heath & Company, 1972), Chapter 23.
Stoker, H. S., and S. L. Seager, *Environmental Chemistry: Air and Water Pollution* (Glenview, Ill.: Scott, Foresman and Company, 1972).
Turk, A. J., and J. T. Wittes, *Ecology, Pollution, Environment* (Philadelphia: W. B. Saunders Company, 1972).

16mm FILMS
Air and Gas Cleaning for Nuclear Energy—Atomic Energy Commission (color, sound, 30 min).
Control of Air Pollution—Du Art Film Laboratories (color, sound, 5 min).
Effects of Air Pollution—Du Art Film Laboratories (color, sound, 5 min).
Poisons, Pests, and People (2 parts)—Contemporary Films (b/w, sound, 30 min).
Radioactive Waste Disposal—Du Art Film Laboratories (color, sound, 24 min).
The Silent Spring of Rachel Carson—Contemporary Films (b/w, sound, 54 min).
Sources of Air Pollution—Du Art Film Laboratories (color, sound, 5 min).
Problems of Conservation: Water—Encyclopedia Britannica Films (color, sound, 16 min).
The Garbage Explosion—Encyclopedia Britannica Films (color, sound, 16 min).

Spectral Analysis

OBJECTIVE
The student shall be able to describe the types of spectral analysis that are in general use, and the types of compounds or elements that can be identified by using each type.

DEMONSTRATION
1. A transparency showing the electromagnetic spectrum should be prepared and displayed during the discussion. Copies of the transparency should be made and distributed to the students for noting important information during the discussion. The regions of the spectrum at which each type of analysis takes place as well as the types of molecules identified in each region should be indicated.

2. Prepare slides of spectral analysis instruments that are used in local laboratories, both industrial and municipal.

3. Prepare individual slides or transparencies of the absorption or emission spectra of elements and compounds to demonstrate the variations that do exist and how interpretations are made.

4. Block diagrams of the machines used, complete with a description of the function of each component, should be shown to the students.

STUDENT AND TEACHER REFERENCES

Dyer, J. R., *Applications of Absorption Spectroscopy of Organic Compounds* (Englewood Cliffs, N.J.: Prentice-Hall, Inc., 1965).

Gordy, W., et al., *Microwave Spectroscopy* (New York: Dover Publications Inc., 1966).

McLafferty, F. W., *Interpretation of Mass Spectra: An Introduction* (Menlo Park, Calif.: W. A. Benjamin, Inc., 1966).

Platt, J. R., et al., *Systematics of the Electronic Spectra of Conjugated Molecules* (New York: John Wiley & Sons, Inc., 1964).

Sonnessa, A. J., *Introduction to Molecular Spectroscopy* (New York: Van Nostrand Reinhold Company, 1966).

Trost, B., *Problems in Spectroscopy: Organic Structure Determination by NMR, IR, UV, and Mass Spectra* (Menlo Park, Calif.: W. A. Benjamin, Inc., 1967).

16mm FILMS

Molecular Spectroscopy—Modern Learning Aids (color, sound, 23 min).
Infrared—Atomic Energy Commission (color, sound, 15 min).
Nuclear Magnet Resonance—Willard Grant Press (color, sound, 28 min).
Analysis by Mass—International Film Bureau (color, sound, 27 min).

8mm FILMS

Absorption Spectra—Ealing Film Loops (color, silent, 4 min).
Visible and Ultraviolet Spectra—Harper & Row, Publishers (color, silent, 4 min).
Infrared Spectroscopy—Harper & Row, Publishers (color, silent, 3 min).
Nuclear Magnetic Resonance—Harper & Row, Publishers (color, silent, 4 min).
Mass Spectrometer—Particles: Relative Abundance—Encyclopedia Britannica (color, silent, 5 min).
Mass Spectrometer—Behavior of Particles in Magnetic and Electric—Encyclopedia Britannica (color, silent, 4 min).

PART...3

Resources for Teaching Physics

Performance Objectives

For each demonstration the student is to observe and then react to what is shown in the demonstration. Although a performance term for this reaction by the students is included for each demonstration, the teacher may consider other performance terms to be more appropriate. Since the qualifying terms for the performance objectives should consider the students' potential and the type of instruction that is provided, these qualifying terms must be provided by the teacher.

Distance Measurement by Triangulation

OBJECTIVE

The student shall be able to set up proportional relationships from similar triangles and use them to compute unknown distances from known and/or measured values.

DEMONSTRATION

The diameter of the sun is measured by triangulation. Cut a hole about 1 cm in diameter in the center of a 4 inch by 6 inch index card. Tape a piece of aluminum foil to the card so that the hole is covered. Make a neat pinhole through the foil at the center of the card. This procedure results in a neater hole and a sharper image than would be possible if the pinhole were made directly through the card. Attach this card to one end of a meter stick with tape or tacks as shown in Figure 3-1. Attach a second card to the other end of the meter stick to serve as a screen.

The meter stick should be held parallel to the sun's rays in bright sunlight so that an image of the sun is formed on the lower card or "screen." **The students should be cautioned not to look directly at the sun.**

The diameter of the sun's image upon the card may be measured directly. Using similar triangles as sketched in Figure 3-2, the diameter of the sun may be computed from the following relationship:

$$\frac{\text{diameter of sun}}{\text{distance from pinhole to sun}} = \frac{\text{diameter of sun's image}}{\text{distance from pinhole to sun's image}}$$

Performance Objectives 51

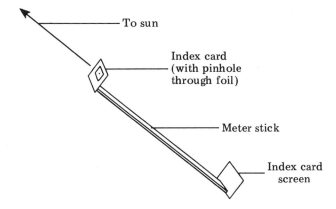

Figure 3-1. Apparatus arrangement.

The pinhole-sun distance is the distance from the earth to the sun, about 1.5×10^{11} m. The students should be able to recognize that the pinhole-image distance is known to be 1 meter. By using an expression such as the equation given, the diameter of the sun can be computed.

This is a good introductory exercise especially if done in pleasant, clear September weather. If it is performed by the students working individually or in teams, their results may be pooled and the class's mean value can be computed. A discussion of the reliability of this mean value versus an individual's values may be productive and the mean deviation or more sophisticated statistical methods may also be introduced. An accepted value for the diameter of the sun is 1.4×10^9 m.

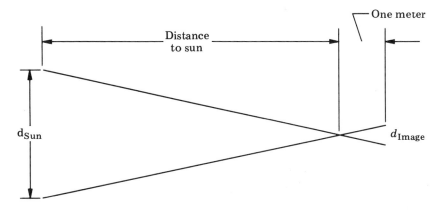

Figure 3-2. Computation diagram; not drawn to scale.

The distance from the earth to the sun is given as a known value in this exercise. Excellent measurements of this distance have been made with radar, but the ancient astronomers' geometrical techniques are more consistent with the spirit and accuracy of this exercise. Interested students may be referred to a discussion of these techniques.

STUDENT REFERENCES
Physical Science Study Committee, *Physics* (Lexington, Mass.: D. C. Heath & Company, 1965), 2d ed., Chapter 3. A general discussion of indirect measurement techniques, including triangulation.
Rogers, Eric, *Physics for the Inquiring Mind* (Princeton, N.J.: Princeton University Press, 1960), Chapter 14. A good treatment of the observation and measurement techniques of the ancient astronomers.

TEACHER REFERENCES
Rutherford, F. J., G. Holton, and F. Watson, Directors, *Project Physics Teacher Resource Book* Unit 1 (New York: Holt, Rinehart and Winston, Inc., 1970). An excellent general reference, even for the teacher who is not using Project Physics. Naked-eye astronomy is experiment E1. Experiment E14 uses a "similar-triangles" computation to measure the height of Piton, a mountain on the moon, using data from a photograph of the moon.

FILMS
Measuring Large Distances—Modern Learning Aids (29 minutes). In this PSSC film, Fletcher Watson describes astronomical-distance measurement using triangulation, parallax, and the inverse square law for light intensity.

Average Velocity and Instantaneous Velocity

OBJECTIVE
The student shall be able to recognize instantaneous velocity to be a limiting case that is obtained when very small time intervals are considered.

DEMONSTRATION
By definition:

$$\text{Average Velocity} = \frac{\text{Change in displacement}}{\text{Time interval required for this change}} = \frac{\Delta d}{\Delta t},$$

and the instantaneous velocity is the limiting value obtained when the time interval in this relationship becomes *very* small. In this demonstration, Δt is reduced by considering smaller and smaller displacements around a given point.

Incline a linear air track so that a glider or cart accelerates down the track when released from rest. Attach a 10 cm high and 30 cm long piece of cardboard, stiff paper, or thin metal to the top of the cart. This card should be mounted in the vertical plane parallel to the cart's axis and should be arranged symmetrically upon the cart. Mount a photocell gate (Ealing A33-9168 or Thornton PHC-100) along the air track so that the card interrupts the light beam when the cart passes through. By connecting this gate to a timer (Ealing A12-4586 or A33-8996, or Thornton DEC-101) the time interval during which the beam is interrupted can be measured. The average velocity during this period can be computed since the change in displacement equals the card's length. By using progressively shorter cards down to 10 cm or even 5 cm in length, a good estimate can be made of the cart's instantaneous velocity at the location of the detector. The track inclination, detector position, initial release position, and initial velocity for the cart should all be held constant throughout data collection.

STUDENT REFERENCES
Physical Science Study Committee, *Physics* (Lexington, Mass.: D. C. Heath & Company, 1965), 2d. ed., Chapter 5.
Rutherford, F. J., G. Holton, and F. Watson, Directors, *The Project Physics Course Text* (New York: Holt, Rinehart and Winston, Inc., 1970), Chapter 1.
(Both of these references include an excellent discussion of average and instantaneous speed. Graphical analysis of motion is discussed.)

TEACHER REFERENCES
Rutherford, F. J., G. Holton, and F. Watson, Directors, *Project Physics Teacher Resource Book* Unit 1 (New York: Holt, Rinehart and Winston, Inc., 1970). Several techniques of demonstrating instantaneous speed are presented in demonstration D3.
Halliday, David, and Robert Resnick, *Physics for Students of Science and Engineering* (New York: John Wiley & Sons, Inc., 1966), Chapter 3. A calculus treatment of instantaneous velocity.

FILMS
Analysis of a Hurdle Race—Part I (film-loop).
Analysis of a Hurdle Race—Part II (film-loop).
These Project Physics film-loops (available from Ealing Corporation) present a hurdle race in slow-motion photography permitting analysis of variations in speed during the race.

Velocity and Acceleration in Linear Motion

OBJECTIVE
The student shall be able to use position-time data from the motion

of an object moving along a straight-line path to graph and to describe the object's velocity and acceleration.

DEMONSTRATION

Any apparatus that marks the position of a moving object at regular time intervals can be used to provide the data. A Behr free-fall apparatus (CENCO 74905), polaroid photography of a moving object using a strobe light for illumination (Ealing A34-2006), the Project Physics bulldozer with felt-tipped pen, or an ISCS water-drop cart (Silver Burdett 5917J55) are all possible sources of this type of data.

The ISCS material may be available at a local junior high school. Alternatively, a thistle-tube shape reservoir holding about 10 ml of water with a *fine* bore tube can be used to provide drops of water at intervals (Silver Burdett 5991J14). If this is attached to a dynamics cart (Macalester 802) or even a motor-driven toy, the motion of this vehicle can be studied using the path of water drops to provide a clearly measurable record of positions at regular time intervals.

Graphical analysis of the data seems to be the most productive. The average velocity at various times can be computed from the data or from the slope of a position-time graph. Similarly, the slope of the velocity-time graph can be used to provide a record of the acceleration.

STUDENT REFERENCES

Rutherford, F. J., G. Holton, and F. Watson, Directors, *The Project Physics Course Text* (New York: Holt, Rinehart and Winston, Inc., 1970), Chapters 1 & 2.

Physical Science Study Committee, *Physics* (Lexington, Mass.: D. C. Heath & Company, 1965), 2d ed., Chapter 5. A detailed description of the technique of graphical analysis of motion.

TEACHER REFERENCES

Rutherford, F. J., G. Holton, and F. Watson, Directors, *Project Physics Teacher Resource Book* Unit 1 (New York: Holt, Rinehart and Winston, Inc., 1970). Experiments E4, E5, and E6 all deal with studies of accelerated motion.

Halliday, David, and Robert Resnick, *Physics* (New York: John Wiley & Sons, Inc., 1966), Chapter 3. This chapter is a sophisticated treatment of "motion in one dimension."

FILMS

Straight Line Kinematics—Modern Learning Aids (34 minutes). In this PSSC film, the one-dimensional motion of an automobile is examined. The displacement, velocity, and acceleration of the car are recorded using black-box instrumentation, and a graphical analysis of the motion is also presented.

Acceleration Due to Gravity—Method I (film-loop).

Acceleration Due to Gravity—Method II (film-loop). These Project Physics film loops distributed by Ealing Corporation present the motion of a falling bowling ball in slow-motion photography. Two different analysis techniques are illustrated.

Velocity and Acceleration in Free-fall—Modern Learning Aids (film-loop). Velocity and acceleration vectors are displayed along with a spot representing the position of an object as it is thrown vertically upward.

One Dimensional Acceleration—Ealing Corporation (film-loop). Stop-action photography of pucks on an inclined air table enables students to compute the acceleration of the pucks as they move across the table.

One-Dimensional Motion—Ealing Corporation (film-loop). Real-time plots of the displacement, velocity, and acceleration are shown for the motion of a cart that is pulled along a track.

Inertia

OBJECTIVE

The student shall be able to analyze situations where inertial effects are important and be able to relate inertia and mass.

DEMONSTRATIONS

A series of demonstrations is suggested. Confronting the student with a variety of phenomena in which it is evident that "objects tend to maintain constant velocity as long as no *net* force is applied" is an effective way of developing this concept whether one follows an inductive or a deductive approach in teaching.

A. Suspend a 1 kg object by a thread or light string and tie an identical piece of thread to the bottom of the object. A suggested procedure to capture the class's attention is to ask, "If I pull down on the lower thread, will the thread break above or below the object?" In response to the answer, "Above the object," a sharp pull on the thread will break the thread below the object; for an answer "Below the object," a light, even pull breaks the thread above the object.

B. The tablecloth-and-china magic trick is another effective demonstration of inertia. A glass of water and a sheet of paper are an easily available alternative for classroom use. This demonstration is familiar to most students, but it is still interesting to watch and several questions can be profitably explored. For example, should the pull on the paper be directed upward, horizontally, or downward? Would lace or linen tablecloths be preferable for this demonstration?

Unsuccessful demonstrations, whether planned or unplanned, may also be explored profitably. Can the students identify the force acting upon the object in a pull directed at too great an angle above the table or in a pull that is *not* the sharp, dramatic jerk that is effectively used by a magician demonstrating with china and tablecloth?

C. Make a stack of checkers about three high on a smooth surface. The bottom checker can be replaced by another without disturbing the rest of the stack by flicking another checker along the table. A direct hit will result in the bottom checker's being replaced by the projected checker. To insure visibility, this has to be done in small groups; therefore, it is a possible homework exercise for the students.

Similarly, a nickel can be dropped into a glass by placing it on a card resting on the top of a glass and striking the card sharply. Practice is necessary to perfect this technique. Either allow the students to find the effective technique themselves or having the teacher fail a few times before establishing a consistent pattern of success helps to identify the possible forces upon the coin.

As mentioned earlier, these demonstrations are relatively familiar to most students. They should be done fairly rapidly and the emphasis should be placed upon discussion of how *inertia* can be related to these phenomena.

STUDENT REFERENCES
Cohen, I. Bernard, *The Birth of a New Physics* (Garden City, New York: Science Study Series, Doubleday & Company, Inc., 1960). This book deals with the development of the mechanics that culminated in Newtonian mechanics. Chapter 5, Toward an Inertial Physics, specifically deals with Galileo's dynamics and the development of the concept of inertia.
Physical Science Study Committee, *Physics* (Lexington, Mass.: D. C. Heath & Company, 1965) 2d ed., Chapter 20.
Rutherford, F. J., G. Holton, and F. Watson, Directors, *The Project Physics Course Text* (New York: Holt, Rinehart and Winston, Inc., 1970), Chapter 3. Like the PSSC text listed, Project Physics presents the concept of inertia in an elegant form.
Stollberg, Robert, and Faith Fitch Hill, *Physics Fundamentals and Frontiers* (Boston: Houghton Mifflin, Company, 1965), Chapter 3. Figure 3-4 uses high speed photography to illustrate the effect of inertia.

TEACHER REFERENCES
Rutherford, F. J., G. Holton, and F. Watson, Directors, *Project Physics Teacher Resource Book* Unit 1 (New York: Holt, Rinehart and Winston, Inc., 1970). The fact that Newton's First Law of Motion is fundamentally a definition is emphasized.

FILMS
Inertia—Modern Learning Aids (26 minutes).
Inertial Mass—Modern Learning Aids (19 minutes).
Inertia develops the relationship that acceleration is directly proportional to force when mass is constant. *Inertial Mass* is a continuation of the other film, dealing with the situation of constant force and variable mass. These are PSSC films.

Inertial Forces—Translational Acceleration—Ealing Corporation (film-loop). A young man is shown riding in an elevator standing while upon a large, sensitive platform balance. A complete trip upward is shown, starting from rest, achieving constant velocity, and then coming to rest at the top; the return trip downward is also shown. The magnitude of the upward force of the elevator is known at all times from the balance reading. This is an excellent film.

Force = mass × acceleration

OBJECTIVE
Given any two of:
1. The resultant of *all* forces acting upon an object.
2. The mass of the object.
3. The object's acceleration,

the student shall be able to compute the third and be able to relate these computations to actual problem situations.

DEMONSTRATION
Making good measurements of both force and acceleration is extremely difficult in demonstration situations. For this reason, the study of film loops might be preferred to actual in-class demonstrations.

STUDENT REFERENCES
Rutherford, F. J., G. Holton, and F. Watson, Directors, *The Project Physics Course Text* (New York: Holt, Rinehart and Winston, Inc., 1970), Chapter 3.
Physical Science Study Committee, *Physics* (Lexington, Mass.: D. C. Heath & Company, 1965), 2d ed., Chapter 20.
(As mentioned earlier, both of these references present an excellent, elegant treatment of mechanics.)
Stollberg, Robert, and Faith Fitch Hill, *Physics Fundamentals and Frontiers* (Boston: Houghton Mifflin, Company, 1965), Chapter 3. The relationship, $F = ma$, is discussed in section 3-8.

TEACHER REFERENCES
Rutherford, F. J., G. Holton, and F. Watson, Directors, *The Project Physics Teacher Resource Book* Unit 1 (New York: Holt, Rinehart and Winston, Inc., 1970). Several accelerometer demonstrations are suggested.

FILMS
Dynamics of Rectilinear Motion—Ealing Corporation (film-loop). The relation, $F = ma$, is investigated using pucks on an air table. Trials are made holding the force constant while different masses are used and with variable force and constant mass. The puck's acceleration can be calculated from the photographs that provide ghost images at equal time intervals.

Newton's First and Second Laws—Ealing Corporation (film-loop). A demonstration of carts on an air track includes a section in which various forces and masses are used. The carts start from rest and accelerate through a measured distance, and the time intervals are measured so that the acceleration can be computed.

Inertial Forces—Translational Acceleration—Ealing Corporation (film-loop). As mentioned earlier, this film shows a student riding in an elevator while standing upon a sensitive balance. His acceleration can be computed from the balance readings.

Newton's Law of Motion—Modern Learning Aids (film-loop). This is a computer-generated film showing the motion of an object with no force acting upon it, with a constant force, and with a changing force acting upon it.

Gravitational Fields and Apparent Weightlessness

OBJECTIVE

The student shall be able to describe the conditions under which apparent weightlessness occurs and be able to identify the forces acting upon the object under these conditions.

DEMONSTRATION

Most of our experience with apparent weightlessness, the condition that exists when apparently no force is required to keep objects from falling under the influence of gravity, has come from the manned space program and the television commentators' descriptions of the conditions that exist in spacecraft. It should be emphasized that the weight, the gravitational attraction of the earth upon the object, is *not* zero. Plotting the force that is exerted upon a 1 kilogram mass due to the gravitational attraction between it and the earth, versus the distance from the center of the earth, is one method of showing this. Another method is simply to examine the inverse square law form of Newton's law of universal gravitation and show that the force cannot be zero as long as the distance is finite.

Once it has been established that the weight of objects in an orbiting spacecraft is not zero, the next step could be an analysis of how an object can appear to be weightless under free-fall conditions in the classroom. A simple demonstration can be done with a large plastic detergent bottle having three or four holes punched in its side. Fill the bottle with water, cap it, then hold it up and allow it to fall into a sink or a bucket. The detergent bottle cap with its hole in the top allows the water to flow out freely through the holes in the bottle when it is held, but minimizes splashing upon impact. If the bottle is held for a moment before it is released, the absence of water flow from the holes *during* free-fall should be evident (Figure 3-3).

A similar demonstration involves attaching a 1 kg mass to a large spring balance (Welch 4091 or CENCO 5150). This should be held up in front of the class and then released and allowed to fall freely for about 2 meters into a *well-padded* receptacle. The dial reading of approximately zero is a graphic illustration of apparent weightlessness.

The analysis for both of these demonstrations is essentially the same as for the astronauts in a spacecraft. The acceleration of the objects (water in the bottle, mass hung from the balance, or a camera in a spacecraft) is the same as the rest of their accelerating, freely falling reference frame (the plastic bottle, the spring balance, or the spacecraft) so that no force is needed to hold the objects in their positions within these accelerating reference frames. Although the weight of the object is not zero, no force need be supplied to counteract this force in these *accelerating*, freely falling reference frames. Hence the television commentator cheerfully describes the situation as one of weightlessness.

STUDENT REFERENCES

McDonald, Robert L., and Hesse, Walter H., *Space Science* (Columbus, Ohio: Charles E. Merrill Publishers, 1970), Chapters 3 and 7. Apparent weightlessness is discussed in Chapter 3, and there is a brief mention of its biological consequences upon humans in Chapter 7.

Rutherford, F. J., G. Holton, and F. Watson, Directors, *The Project Physics Course Text* (New York: Holt, Rinehart and Winston, Inc., 1970), Chapter 3. This chapter includes a good, brief discussion of mass, weight, and gravitation. A thought experiment is included that analyzes the consequences of the floor falling out from under you while you stand upon a bathroom scale.

TEACHER REFERENCES

Gamow, George, *Gravity* (Garden City, N.Y.: Doubleday & Company, Inc., 1960). This book is part of the Science Study Series of paperbacks. It is a rather complete discussion of gravity, including relativistic effects. It is good background reading for the teacher and could be recommended to interested students.

Maleh, Isaac, *Mechanics, Heat and Sound* (Columbus, Ohio: Charles E. Merrill Publishers, 1969), Chapter 5. This paperback contains a more quantitative analysis of weight and weightlessness than its "sister" volume, *Space Science*, which is listed previously.

Dennis, James C., and Larry Choate, "Some Problems with Artificial gravity," *The Physics Teacher*, Vol. 8 (November, 1970), pp. 441-444. An interesting discussion of the problems associated with producing gravity-like effects in spacecraft by causing them to rotate.

Figure 3-3. "Weightlessness" as illustrated by the water in the freely falling bottle at right. Compare the water leakage under the free-fall conditions with the water leakage when the bottle is held stationary as above.

FILM

Inertial Forces—Translational Acceleration—Ealing Corporation (film-loop). A man is shown riding in an elevator while standing upon a sensitive platform scale. Weightlessness is not achieved, but there is a marked decrease in the scale reading while the elevator accelerates downward.

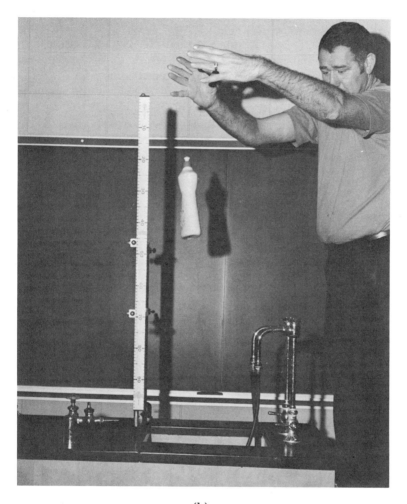

(b)

Newton's Third Law of Motion

OBJECTIVE

The student shall be able to identify the equal and oppositely directed forces that are present when objects interact and the objects upon which each force acts.

DEMONSTRATION

Hook two large, demonstration-type spring balances (Welch 4091 or CENCO 5150) together and have two students pull against each other, using the balances so that their forces can be measured. Ask the students to try to obtain different readings on their balances; for example, ask a strong male student to try to register a greater force upon his balance than that registered upon the balance held by a smaller female student. As precautionary measures, show the students how to pull on the balances so that the spring movements do not bind and request some restraint on the part of the larger student so that the smaller student is not injured. As the students pull on the balances it should become evident that, as closely as can be measured by these scales, the forces are indeed equal in magnitude and oppositely directed.

STUDENT REFERENCES

Rutherford, F. J., G. Holton, and F. Watson, Directors, *The Project Physics Course Text* (New York: Holt, Rinehart and Winston, Inc., 1970), Chapter 3. Newton's statement of this law is presented, both in terms of what it does and what it does not say.

Physical Science Study Committee, *Physics* (Lexington, Mass.: D. C. Heath & Company, 1965) 2d ed., Chapter 22. This chapter includes forces of interaction in the context of the discussion of conservation of momentum.

Stollberg, Robert, and Faith Fitch Hill, *Physics Fundamentals and Frontiers* (Boston: Houghton Mifflin Company, 1965), Chapter 3. Newton's Third Law of motion and its relationship to force and momentum is discussed in sections 3-21 and 3-22.

Rogers, Eric, *Physics for the Inquiring Mind* (Princeton, N.J.: Princeton University Press, 1960), Chapter 7. Action and reaction is thoroughly discussed by Rogers. Good, thought-provoking problems are included.

TEACHER REFERENCES

Rutherford, F. J., G. Holton, and F. Watson, Directors, *Project Physics Teacher Resource Book* Unit 1 (New York: Holt, Rinehart and Winston, Inc., 1970). Includes a discussion of the Project Physics transparency, No. 8, The Tractor-Log Paradox, which can be utilized with or without the commercial transparency.

FILMS

Newton's Third Law—Ealing Corporation (film-loop). This is a good film loop using an air track. Two carts are initially at rest with a compressed spring between them. Their respective accelerations after release can be determined from their displacements in equal time intervals. The last part of the film shows the oscillation of pairs of carts when they are joined together by a spring.

Motion Through Space

OBJECTIVE

The student shall be able to analyze motion in two or three dimensions in terms of orthogonal components.

DEMONSTRATION

The "monkey-gun" demonstration is based upon a common textbook problem: A monkey hanging from a branch drops from the tree at the instant the hunter fires at him. If the hunter has aimed directly at the monkey, the shot will hit its mark as both the target and the projectile experience the same vertical acceleration due to gravity.

The apparatus is available commercially (CENCO 75412) or it can be constructed using one-half inch rigid copper tubing (or similar material). A metal can is normally used as the target, but once the gun is sighted in, very impressive results are obtained by the use of clay-pigeon targets. By taping a small metal tab to the clay pigeon, the target can be held to the electromagnet until the circuit is broken as the projectile leaves the gun.

The arrangement as shown in Figure 3-4, with the gun aimed *horizontally* at the target, provides a relatively simple case for anal-

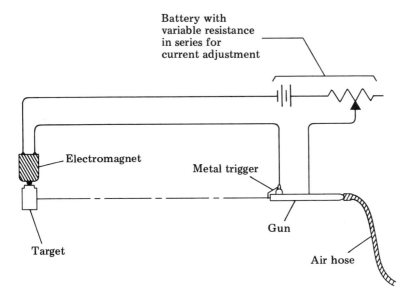

Figure 3-4. Apparatus arrangement for monkey-gun demonstration; note how the gun is bore-sighted at the target.

ysis. In this arrangement, the projectile and the target start with the same vertical displacement and velocity and have the same vertical acceleration, 9.8 meters/second/second. Since the projectile and target will always have the same vertical displacements under these conditions, the reason why the target is hit should be evident. By measuring the target's displacement at the time of impact, the time of flight can be determined; if the horizontal displacement is measured, the projectile's muzzle velocity can be computed.

Note that horizontal mounting of the gun may require clamping the gun to ceiling pipes or to a special bracket if a relatively large vertical displacement is desired. If both the magnet and the gun must be mounted, horizontal mounting of the gun should present no serious problem. In either case, it is well worth the effort to arrange the horizontal mounting since the analysis is considerably simplified by this configuration.

STUDENT REFERENCES

Holton, Gerald, and Duane Roller, *Foundations of Modern Physical Science* (Reading, Mass.: Addison-Wesley Publishing Co., Inc., 1958), Chapter 3. Although the computations are considerably more complicated than the situation presented in this demonstration, the relationship between projectile range and angle of fire is developed.

Physical Science Study Committee, *Physics* (Lexington, Mass.: D. C. Heath & Company, 1965), 2d ed., Chapter 20.

Rutherford, F. J., G. Holton, and F. Watson, Directors, *The Project Physics Course Text* (New York: Holt, Rinehart and Winston, Inc., 1970), Chapter 4.

(Both PSSC and Project Physics present a thorough discussion of projectile motion.)

Stollberg, Robert, and Faith Fitch Hill, *Physics Fundamentals and Frontiers* Boston: Houghton Mifflin Company, 1965), Chapter 2. Only projectiles with horizontal or vertical initial velocities are considered.

TEACHER REFERENCES

Rutherford, F. J., G. Holton, and F. Watson, Directors, *Project Physics Teacher Resource Book* Unit 1 (New York: Holt, Rinehart and Winston, Inc., 1970).

Halliday, David, and Robert Resnick, *Physics* (New York: Holt, Rinehart and Winston, Inc., 1970), Chapter 4. As in most college physics texts, the general situation of any angle of inclination for the gun is discussed. An example specifically analyzes this type of "monkey-gun" demonstration.

FILMS

Free Fall and Projectile Motion—Modern Learning Aids (27 minutes). Along with a study of freely falling bodies, a slow-motion presentation of a large monkey-gun demonstration is shown in this PSSC film.

Velocity and Acceleration in Free Fall—Modern Learning Aids (film-loop). This film is an animated representation of the motion of an object that is thrown vertically upward. Velocity and acceleration vectors are displayed along with the spot that indicates the position of the object.

Trajectories—Ealing Corporation (film-loop). Different projectile trajectories are demonstrated using pucks on a tilted air table. Data is collected for determining the angle of projection for maximum horizontal range.

Vectors

OBJECTIVE

The student shall be able to resolve a single vector into its orthogonal components and to compute the resultant of two or more vectors.

DEMONSTRATION

Using loops of wire or strong cord, attach two demonstration balances (Welch 4091 or CENCO 5150) to a horizontal support rod as shown in Figure 3-5. At least one of these loops should be movable so that the angle between the two forces can be varied. A strong cord or fish line should be used between the two balances to support the load, m. The range of the balances used determined the appropriate value for m, but a 1 kg mass works well with 0-20 Newton balances.

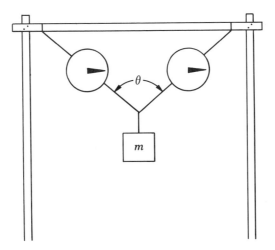

Figure 3-5. Apparatus arrangement for the vector demonstration.

When taking balance readings, both balances should read the same when the system is arranged symmetrically. This can be easily adjusted if the load m is hung from a loop tied around the string between the two balances. The relationship between the tension in the string and θ, the angle between two forces, can be established by collecting data for the forces at various angles and plotting force versus the angle θ.

Using an overhead projector and a transparency on which the coordinate axes are drawn permits the data to be easily and accurately plotted. The theoretical curve for the relationship between the force and the angle between the cords can be drawn upon another transparency that can then be laid on top of the experimental data to show the fit between the data and the theoretical curve. Since the system is in equilibrium

$$2F \cos(\theta/2) = mg,$$

the theoretical curve can be drawn using the relation,

$$F = mg/[2 \cos(\theta/2)].$$

The following questions can be asked about this curve:
What does the force represent when the angle θ is $0°$?
What value does the force approach as the angle θ approaches $180°$?

STUDENT REFERENCES
Project Physics, *Vectors*, program (New York: Holt, Rinehart and Winston, Inc., 1970). This program, which was produced by Project Physics, could be used for self-instructional purposes by very capable students or as a supplement to regular class instruction for students who need extra work with vectors.

Crane, H. R., *Programmed Math Reviews* (New York: New Century, 1966). This is a collection of five programmed reviews one of which deals with vectors. The other reviews deal with the use of exponents, algebra, triangles and simple trigonometry.

Williams, John E., H. C. Metcalfe, F. E. Trinklein, and R. W. Lefler, *Modern Physics* (New York: Holt, Rinehart and Winston, Inc., 1968), Chapter 3. Although using only forces as examples of vector quantities, solutions using the law of sines and the law of cosines are illustrated.

TEACHER REFERENCES
Rogers, Eric M., *Physics for the Inquiring Mind* (Princeton, N.J.: Princeton University Press, 1960), Chapter 2. Many sample problems with vectors are included. This approach, and indeed the chapter, could be recommended to students who do not care for the programmed materials.

FILMS

Vector Addition I—Velocity of a Boat—Ealing Corporation (film-loop). This Project Physics film shows an overhead view of a boat in a river. Vector addition of the boat and stream velocities is shown as the boat follows different courses relative to the current.

Vector Kinematics—Modern Learning Aids (16 minutes). The acceleration and velocity vectors are shown for various types of motion. The vectors are computed by a computer and displayed on a cathode ray tube screen. This is a PSSC film.

Conservation of Momentum

OBJECTIVE

The student shall be able to describe the motion of a rocket in terms of the momentum of the entire rocket-plus-exhaust-gas system.

DEMONSTRATION

Inflate a large balloon, hold its opening closed, and then release the balloon horizontally. The motion of the balloon will be rapid and erratic, and easily visible. The question, "What pushes the balloon forward?" can be used to lead the class into a discussion of the system of the balloon and its contents.

Both for a rocket and for an inflated balloon that are initially at rest in a frame of reference, the total momentum is initially zero. In this frame of reference the exhaust gas from a rocket or the air escaping from a balloon has a nonzero momentum. However, if no net external force acts upon the system, its momentum must remain constant and equal to zero. Assuming that the air drag is negligible, at least at the time of release, this requirement is satisfied for the horizontal motion. Therefore, the rocket or balloon must acquire a net horizontal momentum that is equal and opposite to that of the expelled exhaust gases. The common misconception that the exhaust gases must push against something outside of the rocket, such as the ground or the air, should be brought into the discussion and shown to be inaccurate. It is useful to point out that rockets work very well in the very high vacuum of outer space.

STUDENT REFERENCES

Physical Science Study Committee, *Physics* (Lexington, Mass.: D. C. Heath & Company, 1965), 2d ed., Chapter 22. This chapter, momentum and the conservation of momentum, provides an excellent analysis of momentum. Several excellent photographs are included. The last part of the chapter deals specifically with rockets and the conservation of momentum.

Stollberg, Robert, and Faith Fitch Hill, *Physics Fundamentals and Frontiers* (Boston: Houghton Mifflin Company, 1965), Chapter 3. Rocket motion is considered along with Newton's third law. The actual operation of the rocket engine is discussed in sections 12-19.

TEACHER REFERENCES

Halliday, David, and Robert Resnick, *Physics* (New York: John Wiley & Sons, Inc., 1966), Chapter 9. This is a sophisticated treatment using vector calculus.

Gottlieb, Herbert H., "Conservation of Momentum in an Explosion," *The Physics Teacher*, 8:90 (February, 1970). A description of an experiment involving the use of two PSSC dynamics carts to illustrate the conservation of momentum in an explosion. No timing apparatus is required.

FILMS

Moving with the Center of Mass—Modern Learning Aids (26 minutes). This PSSC film uses dry ice pucks with attached magnets to illustrate elastic interactions. Conservation of energy and momentum are illustrated.

Conservation of Momentum: Elastic Collisions—Ealing Corporation (film-loop).

Conservation of Momentum: Inelastic Collisions—Ealing Corporation (film-loop).

(Both of these films deal with collisions between gliders on an air track. Data may be collected from these films that permit the computation of the system's momentum after the collisions.)

Recoil—Ealing Corporation (film-loop). This Project Physics film shows a gun firing a bullet. Direct measurements can be made of the speeds of the bullet and the recoiling gun.

Energy Transformations and Conservation of Energy

OBJECTIVE

The student shall be able to analyze the behavior of a simple pendulum in terms of energy transfer between gravitational potential energy and kinetic energy, and to discuss the effects of the loss of energy from this system.

DEMONSTRATION

From a *secure* ceiling support, suspend a large pendulum bob using about 3m of strong fishing line or nylon rope. The mass of the bob should be at least several kilograms; an old bowling ball works very nicely. The pendulum should swing through a large arc that is visible to the class.

An effective way to begin this demonstration is to be involved in hanging the pendulum as the students enter the classroom. With the pendulum bob held at one end of its arc, discuss the energy conditions that exist at this point when the bob is at rest. Compare and contrast these energy conditions to those at the lower-most point of the arc and at the top of the arc on the opposite side. From a consideration of gravitational potential energy and energy conservation, the class should be able to agree that the pendulum will not rise beyond the original height from which it was released from rest. However, if energy is conserved, the pendulum bob should return to this height at the far end of its arc and when it returns to its initial position.

If the class is certain that the bob will not rise beyond this initial height, a good test is to mark the initial position and compare it with the position of the bob after one oscillation. Either the teacher or a student volunteer can stand at the end of the arc, hold the bob up to the end of his nose, and release it from rest. Even a student who is convinced that the bob will not rise past its initial position will probably flinch as the bob returns back up toward his face for the first time. All knots and anchorages must be secure, since a considerable strain may be placed upon them by the heavy, swinging pendulum bob.

Once the pendulum bob's initial position is marked clearly, it becomes apparent that the bob does not even return to its initial position. Therefore, the total energy of the system, that is, its gravitational potential energy plus its kinetic energy must be decreasing. A discussion of this energy loss can be profitable.

STUDENT REFERENCES

Rutherford, F. J., G. Holton, and F. Watson, Directors, *The Project Physics Course Text* (New York: Holt, Rinehart and Winston, Inc., 1970), Chapter 10. Conservation of mechanical energy, the topic of this demonstration, is considered in section 10-3.

Shockley, William, and Walter A. Gong, *Mechanics* (Columbus, Ohio: Charles E. Merrill Publishers, 1966). This book is a paperback in the Merrill physical science series. Conservation principles are discussed at considerable length. The simple pendulum is discussed in terms of energy conservation in Chapter 9.

Physical Science Study Committee, *Physics* (Lexington, Mass.: D. C. Heath & Company, 1965), 2d ed., Chapter 24. Graphical analysis of energy exchanges is discussed.

TEACHER REFERENCES

Rutherford, F. J., G. Holton, and F. Watson, Directors, *Project Physics Teacher Resource Book* Unit 3 (New York: Holt, Rinehart and Winston, Inc., 1970).

FILMS

Conservation of Energy—Aircraft Takeoff—Ealing Corporation (film-loop). An airplane takes off with constant applied power. The kinetic and potential

70 Resources for Teaching Physics

energies of the airplane can be measured at three different points; in level flight before climbing, during a climb, and during level flight at the greater altitude.

Conservation of Energy—Pole Vault—Ealing Corporation (film-loop). In this Project Physics film, the total energy of a pole vaulter is measured as he goes up and over the bar. Energy transformations between kinetic, elastic potential, and gravitational potential forms are considered.

Conservation of Energy—The Pendulum and the Piledriver—Ealing Corporation (film-loop). The student is able to compare systems in which mechanical energy is conserved quite well, the simple pendulum, with a system in which there is a considerable loss of mechanical energy, the piledriver. Quantitative data may be collected for the pendulum.

Energy Transfer—Modern Learning Aids (film-loop). Two long, simple pendulums coupled together are shown in motion. The variation in their amplitudes with time is indicated by the trail of sand released from the bobs. The effect of changing the coupling is also shown. This film could be used effectively as a device to begin a discussion of energy transfers.

Conservation of Energy and Projectile Motion

OBJECTIVE

The student shall be able to predict and/or quantitatively interpret the impact points of streams of water from various apertures of a water-filled container (Figure 3-6).

DEMONSTRATION

There are many incorrect descriptions of the water can paradox. Biser discusses several of these in his article listed in the teacher references. The most common error seems to be to overemphasize the increase in pressure with depth, and consequently the velocity at which the water is ejected. However, the horizontal range to point of impact depends upon the time of flight as well as the horizontal velocity.

It is a straightforward problem in differential calculus to show that the range to a point of impact on a plane at the base of the water-filled container is a maximum for an aperture located midway between the top surface of the water and the base of the container. The same result can also be demonstrated by graphical analysis. The time of flight can be computed from $y = \frac{1}{2}gt^2$, where y is the height of the aperture above the base of the container. The velocity with which the water is ejected from the hole can be obtained from a consideration of the energy conditions:

Figure 3-6. A water-can demonstration illustrating that the stream from the hole at ½h has the maximum range.

| Potential energy that would be lost by a mass m of water in descending to the aperture | = | Kinetic energy with which the mass of water, m, is ejected |

$$mgh - mgy = \tfrac{1}{2} mv^2$$

where y is the height of the aperture above the plane at the base of the container and h is the height of the top surface of the water above the base of the container. Therefore, the velocity, v, at which the water leaves the aperture is given by the equation:

$$v = \sqrt{2g(h-y)}.$$

The horizontal range to the point of impact of the water on a plane at the base of the container of water is:

$$\text{Horizontal Range} = vt = 2\sqrt{(h-y)y}$$

$$\text{Since } t = \sqrt{\frac{2y}{g}}$$

A graph of horizontal range versus y, the height of the aperture, will clearly show a maximum when $y = \tfrac{1}{2}h$, where h is the height of the top surface of the water above the base.

A 5-gallon paint container or similar can works well as an apparatus for this demonstration. Instead of having the students "confirm" a theoretical argument such as that given, it might be productive to give them one of the incorrect descriptions in the literature and textbooks as a model to be tested, without any further information. This would produce a controversy over what to believe, the experiment or the book.

STUDENT REFERENCES
White, Harvey E., *Descriptive College Physics*, 2d ed. (New York: Van Nostrand Reinhold Company, 1963). A nonmathematical college text that can be comprehended by high school physics students. A good illustration and discussion of the trajectories of streams of water is included.

TEACHER REFERENCES
Biser, Roy H., "The Water Can Explored Again," *The Physics Teacher, 4:* 304–305 (1966). Sixteen illustrations of the trajectories of streams of water, most of them inaccurate, are presented along with their sources.

Paldy, Lester G., "The Water Can Paradox," *The Physics Teacher, 1:* 126 (1963). A discussion and a derivation of the relationship between aperture height and horizontal range.

Uniform Circular Motion

OBJECTIVE
The student shall be able to identify the source of the centripetal force acting upon an object that is moving along a circular path and indicate how this force must be changed in order to maintain the same path radius when the object's speed is changed.

DEMONSTRATION
A good device in which the source of the centripetal force is clearly evident is the familiar demonstration of swinging a pail partially filled with water in a circular path in a vertical plane. Be sure that

you have adequate clearance; watch out for overhead lights, front-row desks, and lecture table. A little practice may be necessary before you can start and stop the movement of the pail without spilling any water. If there are some members of the class in line with the pail's path, they will emphasize the necessity of maintaining the centripetal force if you say, "Now if I let go of the pail..." while swinging it.

More quantitative measurements can be made with the simple apparatus described in the references, but these are more suited to laboratory investigation.

STUDENT REFERENCES

Physical Science Study Committee, 2d ed., *Laboratory Guide Physics* (Lexington, Mass.: D. C. Heath & Company, 1965). Experiment III-5. This is an experiment in which the elements of centripetal force are investigated. The only apparatus required is a timer, about 1.5 m of thin, strong cord, a two-hole rubber stopper, a short piece of glass tubing, and some weights.

TEACHER REFERENCES

Rutherford, F. J., G. Holton, and F. Watson, Directors, *Project Physics Teacher Resource Book* Unit 1 (New York: Holt, Rinehart and Winston, Inc., 1970). Experiment E12 describes the same experiment as the one listed in the PSSC laboratory manual. Chapter 4 in the *Project Physics Text* unit, *Concepts of Motion*, includes circular motion and centripetal acceleration, along with a discussion of satellite orbits.

FILMS

Inertial Forces: Centripetal Acceleration—Ealing Corporation (film-loop). The effects upon the passengers in an amusement park Rotor-Ride are shown from both inside and outside of this rotating frame of reference.

Uniform Circular Motion—McGraw-Hill Films (8 minutes). Circular motion is defined and demonstrated along with centripetal force. Several examples of uniform circular motion are shown.

Circular Motion—Ealing Corporation (film-loop). A puck tied to a string moves along a circular path on an air table. From the ghost images of the puck, the speed of the puck along the path can be compared with the tangential speed once the string is removed. An accelerometer is used to confirm the acceleration direction obtained from a consideration of instantaneous velocity vectors.

Temperature

OBJECTIVE

The student shall be able to predict how the height of the liquid in a Galilean thermometer will change when the temperature of the confined gas is changed.

DEMONSTRATION

The arrangement of the apparatus is shown in Figure 3-7. An air thermometer tube (CENCO 77300) can be substituted for the round bottom flask and glass tubing. Fill the gas-collecting bottle about one-third full of a water and food coloring mixture. Slide the two rubber stoppers on the glass tubing so that they are positioned as shown in the figure and then insert the one-hole stopper into the mouth of the flask. Invert the flask and insert the glass tubing down into the reservoir of water and food coloring mixture. Adjust the amount of air in the bulb so that the liquid level is clearly visible and about halfway between the liquid level in the reservoir and the top of the glass tube when the air in the top bulb is at room temperature.

A class vote on whether the liquid column will go up, go down, or stay in place when the top flask is heated is a possible way to begin the discussion. The flask may be heated by holding it in your hand. You should have an ice cube available so that a hypothesis formulated by the students on the basis of their observations of the gas-filled bulb being warmed can be tested by cooling the gas-filled bulb.

The analysis of this simple demonstration can be made increasingly sophisticated. For example, an analysis can be made of how this

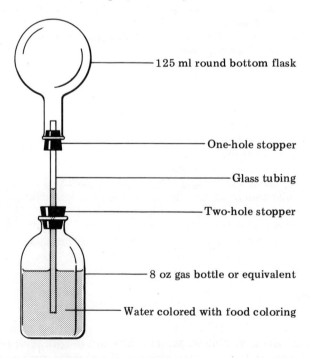

Figure 3-7. Apparatus arrangement for the Galilean thermometer.

simple temperature measuring device could lead to the invention of the barometer. The students can also estimate or even compute the pressure upon the gas inside the bulb at each of the observed temperatures. The influence of vapor pressure could also be examined: note the condensation inside the bulb when it is cooled. However, the basic purpose of this demonstration is to introduce an operational means of measuring temperature that is similar to the procedures used in all liquid or gas expansion thermometers.

STUDENT REFERENCES

Maleh, Isaac, *Mechanics, Heat and Sound* (Columbus, Ohio: Charles E. Merrill Publishers, 1969), Chapter 7. This paperback in the Merrill physical science series includes a discussion of thermometry and the Galilean thermometer and its shortcomings.

Stollberg, Robert, and Faith Fitch Hill, *Physics Fundamentals and Frontiers* (Boston: Houghton Mifflin Company, 1965). Chapter II. A Galilean thermometer is shown in figure II-1.

TEACHER REFERENCES

Rutherford, F. J., G. Holton, and F. Watson, Directors, *Project Physics Teacher Resource Book* Unit 3 (New York: Holt, Rinehart and Winston, Inc., 1970). Experiment E26 is a good experiment dealing with thermometers and temperature measurement.

Rogers, Eric, *Physics for the Inquiring Mind* (Princeton, N.J.: Princeton University Press, 1960), Chapter 27. The general concept of temperature is discussed in an elementary but detailed fashion.

FILMS

Thermal Expansion of Gases—Ealing Corporation (film-loop). A Galilean thermometer is demonstrated showing that gases expand when they are heated. Given the assumption that all gases undergo expansion at the same rate when heated, the student is provided with an opportunity to use the data to measure temperatures.

Heat and Cold: Basic Theory of Heat, Part II (15 minutes). CENCO film, available also from Bureau of Audio-Visual Instruction, University of Wisconsin. Heat transfer and the measurement of heat is discussed along with temperature scales.

Heat and Work

OBJECTIVE

The student shall be able to compute the temperature increase that would be expected if a known quantity of heat, determined from the amount of work done upon the system, were added to a sample of metal shot of known mass and composition, and to predict whether the *actual* temperature change in an experimental test of this situation would be greater than, equal to, or less than the computed temperature change.

DEMONSTRATION

The experiment consists of repeatedly lifting a sample of lead shot and allowing it to fall through a distance of 1 meter each time. The mechanical work done upon the shot during this operation can be computed. Assuming that all of this energy goes into heating the shot, an expected temperature increase can be computed. This demonstration can also be used to estimate the mechanical equivalent of heat; however, it seems more productive to use it as the basis of an analysis of energy conversion and transfer.

A mailing tube about 3 inches in diameter and slightly more than 1 meter in length is the basic apparatus. Find or make wooden plugs to fit the ends of the tube and bore a 3/4 inch hole through the center of one plug. The tube should be cut to length so that the shot will fall through a distance of one meter each time. This means that the inside distance between the plugs should be about 101 cm to compensate for the size of the sample of shot. The plugs should be securely fastened in the tube ends with screws, glue, and/or tape. A thermometer should be inserted through a one-hole, number 7 cork or stopper so that the bulb just protrudes past the end of the stopper. The only other apparatus needed is another solid number 7 cork for a plug in the end hole and one-half kilogram of lead shot.

For the demonstration, the tube and lead shot both start out at room temperature. Pour the shot into the tube (a funnel is helpful) and insert the solid stopper into the hole. Grasp the tube at each end, keeping one finger over the stopper in the end plug, and hold the tube vertically. Now rapidly rotate the tube $180°$ so that the shot is lifted to the top and then falls (1 meter) to the lower end of the tube; repeat 100 times. Remove the solid stopper and insert the one bearing the thermometer and allow the shot to flow down around the thermometer bulb and measure the final temperature.

It would be more effective if this experimentally measured temperature were used to test student hypotheses. Therefore the following sequence is suggested:

1. Explain the apparatus' operation to the students.
2. Analyze the energy situation,

    Work Done = Heat Gained

    $$(N)(mgl) = (m)(SH)(\Delta \text{ Temp})(4.2 \times 10^3 \frac{J}{Kcal}),$$

Where:

N = number of rotations = 100
m = .50 kilogram
L = 1.0 meter

SH = Specific Heat of lead
= $.031 \frac{KCal}{Kg\ C°}$

The given values yield a value of 7.5 C° for Δ Temp.

3. Therefore, the final temperature can be predicted from the initial, room-temperature value.
4. The students should be asked to analyze the situation and the assumptions that were made. In particular, the assumption that *all* of the work done upon the shot appears as heat gained by the shot should be discussed. Is the assumption of no heat loss to the environment tenable?
5. On the basis of this analysis, it should become evident that the situation is:

$$T_{room} < \frac{\text{Measured}}{T_{final}} < T_{room} + \Delta \text{Temp.}$$

6. This hypothesis can be tested by the demonstration.

STUDENT REFERENCES
Rutherford, F. J., G. Holton, and F. Watson, Directors, *The Project Physics Course Text* (New York: Holt, Rinehart and Winston, Inc., 1970), Chapter 10. This chapter on energy includes an extensive discussion of the problem of conversion from mechanical energy to heat energy and the early engines designed to convert heat to mechanical work.

TEACHER REFERENCES
Rogers, Eric, *Physics for the Inquiring Mind* (Princeton, N.J.: Princeton University Press, 1960), Chapter 29. This chapter includes a discussion of the historical development of the theory of the relationship between mechanical energy and heat.

FILMS
Mechanical Energy and Thermal Energy—Modern Learning Aids (22 minutes). In this PSSC physics film, several models are presented to illustrate the energy of the bulk motion of matter and the random, thermal motion on a molecular scale. A model is shown in which a dry ice puck interacts with small steel balls so that the macroscopic energy of the puck is converted to the "thermal" energy of the random motion of the balls.

Wave Motion

OBJECTIVE
The student shall be able to state operational definitions in his own words and cite examples for: longitudinal waves, transverse waves, and superposition of waves.

DEMONSTRATION
A long brass spring (CENCO 84740 or Welch 3340) provides an excellent means to introduce transverse wave phenomena. If one end of the spring is securely clamped, a short segment about 1 or 2 meters long can be oscillated in a vertical plane.

If the spring is laid horizontally on a smooth floor and extended to its full length, similar demonstrations may be performed. If a student at the other end also supplies transverse pulses, you can, with a little practice, have wave pulses traveling in opposite directions along the spring. Despite considerable energy losses because of friction, the independent transmission of pulses along the spring is quite evident. Superposition of pulses can also be demonstrated, but some stop-action photos or sketches are helpful as the rapid motion of the pulses makes their superimposed pattern rather difficult to see.

A Slinky (Welch 3339 or local toy stores) can be used in a similar fashion to demonstrate longitudinal waves. The compressions and rarefactions are visible if pulses are repeatedly applied longitudinally, parallel to the axis of the spring.

STUDENT REFERENCES
Physical Science Study Committee, *Physics*, 2d ed. (Lexington, Mass.: D. C. Heath & Company, 1965), Chapter 15. PSSC is famous for its treatment of wave motion, and this chapter is a fine example of their approach. Several excellent series of photographs illustrating the type of waves and wave phenomena produced in the demonstration are included in this chapter. Superposition is discussed and illustrated.

Rutherford, F. J., G. Holton, and F. Watson, Directors, *The Project Physics Course Text* (New York: Holt, Rinehart and Winston, Inc., 1970), Chapter 12. A concise introduction to wave phenomena that proceeds rapidly to a general discussion of interference, reflection, and refraction.

TEACHER REFERENCES
Rutherford, F. J., G. Holton, and F. Watson, Directors, *Project Physics Teacher Resource Book* Unit 3 (New York: Holt, Rinehart and Winston, Inc., 1970). Several introductory wave mechanics demonstrations are described in detail. Demonstrations D37, wave propagation, and D39, superposition, are particularly relevant here.

FILMS

Superposition of Pulses—Ealing Corporation (film-loop). Because of special lighting and photographic techniques, the Ealing film loops of ripple-tank phenomena produce better viewing conditions than are possible in class demonstrations. The crossing of two single pulses is shown and the paths of the intersection points are marked.

Simple Waves—Modern Learning Aids (27 minutes). The propagation of waves is demonstrated by the use of ropes and slinkies. Slow-motion photography is used for some of the demonstrations in this PSSC film.

Progressive Waves: Transverse and Longitudinal—McGraw-Hill (9 minutes). Available from the Bureau of Audio-Visual Instruction, University of Wisconsin. Wave propagation in ropes is illustrated and the dependence of the wave speed upon elastic and inertial factors is discussed. Wave length and frequency are identified for both transverse and longitudinal wave patterns.

Reflection

OBJECTIVE

The student shall be able to predict the patterns resulting from the reflection of waves from straight and circular barriers.

DEMONSTRATION

The reflection of waves in a flipped spring at either a fixed or a free end can be demonstrated using a long brass spring. This is most easily done on the floor.

Reflection of light with mirrors can be effectively demonstrated using blackboard optics materials (Klinger K041000). Similar results can be obtained with a slide projector light source and hand-held mirrors for reflection patterns that are visible upon the blackboard. Light rays can be produced by taping strips of aluminum foil across a standard slide mount so that the opening is covered except for one or more slits about ½mm wide. The slits should be horizontal and the beam from the projector should be almost parallel to the blackboard. Any plane mirrors can be used for these demonstrations (Figure 3-8). Reflection from concave or convex mirrors can be demonstrated with commercially available cylindrical mirrors. (Ealing A22-4659) or even with homemade ones cut from *shiny* surfaced cans (Figure 3-9).

The reflection of other kinds of waves should also be shown. Ripple tank demonstrations are good, although the viewing conditions are probably superior in the Ealing film loops of these demonstrations.

Figure 3-8. Blackboard demonstration of reflection from a plane mirror using a single slit in a slide projector as a light source and a plane mirror set at an angle.

STUDENT REFERENCES
Physical Science Study Committee, *Physics*, 2d ed. (Lexington, Mass.: D. C. Heath & Company, 1965), Chapter 12. A thorough discussion of "reflection and images." Some excellent, thought-provoking problems are included.

Griffin, Donald R., *Echoes of Bats and Men* (Garden City, N.Y.: Doubleday-Anchor Books, 1959). This science-study series paperback discusses the application of reflection of waves to radar distance measurement.

White, Harvey E., *Modern College Physics*, 5th ed. (New York: Van Nostrand Reinhold, 1966). This or any other of Harvey White's general texts can be used as a traditional reference to provide a basic discussion of optics. Chapter 37 of this edition includes a discussion of plane and spherical mirrors and ray tracing.

TEACHER REFERENCES
Jenkins, Francis A., and Harvey E. White, *Fundamentals of Optics* (New York: McGraw-Hill Book Company, Inc., 1957). This is one of the most widely accepted college optics texts, and includes a thorough discussion of reflection and geometrical optics in general.

Figure 3-9. Blackboard demonstration of reflection from a concave mirror made from a three-pound coffee can and using a triple slit in the slide projector as a light source.

FILMS

Straight Wave Reflection from Straight Barriers (film-loop).
Circular Wave Reflection from Various Barriers (film-loop).
Reflection of Waves from Concave Barriers (film-loop).
(These three film loops from the Ealing Corporation contain excellent photography of water-wave reflection in ripple tanks.)

Refraction

OBJECTIVE

The student shall be able to predict whether the direction of motion of a wavefront will be changed toward or away from the normal to the intermedium surface, if given the relative wavespeeds in the two media. The student shall also be able to diagram the paths of light rays as a result of refraction by thin lenses.

DEMONSTRATION

Refraction of light waves can be shown with an optical disk (Ealing A22-5003 or CENCO 85245) or with a blackboard optics set (Klinger K04100). With this apparatus, the light rays can be seen inside the glass or plastic pieces. Begin the demonstration with a rectangular block to show the basic effects of refraction at a plane surface. A light ray from a single slit in a slide projector and a water-filled glass or plastic container with flat, parallel sides shows the same effect (Figure 3-10).

Next, use a prism and point out how the direction of the light ray is changed by refraction as it enters and leaves the prism. Two prisms placed "base to base" crudely approximate a convex lens. Similarly, two prisms placed "point to point" approximate a concave lens. Refraction with thin lenses can be done next or can be studied as a laboratory exercise.

The Ealing film loop, *Refraction of Waves*, shows the refraction of water waves very clearly. The basic idea of refraction can also be presented with the analogy of marchers who change the rate of their steps as they cross a boundary.

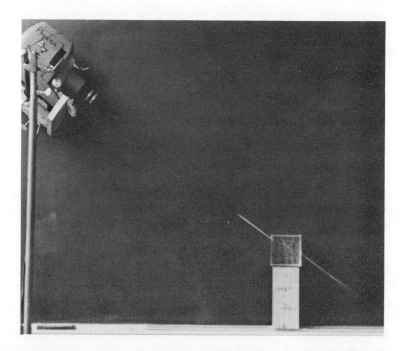

Figure 3-10. Blackboard demonstration of refraction as light from a single slit in the slide projector passes through a 10 cm square plastic box filled with water. Note the reflection.

STUDENT REFERENCES

Physical Science Study Committee, *Physics*, 2d ed. (Lexington, Mass.: D. C. Heath & Company, 1965), Chapter 13. This chapter is a comprehensive treatment of refraction. The effects of varying the angle of incidence and of using various materials are discussed.

White, Harvey E., *Modern College Physics*, 5th ed. (New York: Van Nostrand Reinhold Company, 1966), Chapters 28 and 40. This is a basic treatment of refraction. Optical instruments are discussed in Chapter 41.

TEACHER REFERENCES

Physical Science Study Committee, *Laboratory Guide*, 2d ed. (Lexington, Mass.: D. C. Heath & Company, 1965). Experiment II-5 could be adapted for a demonstration. To illustrate the "refraction" of particles whose speed changes as they cross a boundary.

FILM

Refraction of Waves—Ealing Corporation (film-loop). A ripple tank with a glass plate in it is used to demonstrate the refraction of water waves as they move from deep to shallow water. The critical angle and total reflection are also demonstrated.

Interference

OBJECTIVE

The student shall be able to describe the conditions necessary for a stable interference pattern and predict the situation, such as a maxima or a minima, at specified locations in an interference pattern.

DEMONSTRATION

Ripple tanks or the Ealing film loops of water wave interference patterns serve as a good introduction to the topic of interference since both the waves and the results of their interference are clearly visible.

Interference between sound waves can be demonstrated using a pair of large tuning forks (Welch 3246 or CENCO 84570). The frequency of one of the forks can be varied slightly with a movable weight, and clearly audible beats can be produced. A microphone can be connected to an oscilloscope to provide a simultaneous visual display of these variations in sound amplitude.

Inexpensive replica gratings (CENCO 86252 or Edmund P-30,282) can be purchased in quantity and given to the individual students for observation. Quantitative analysis of line spectra can be done with these gratings, but the intention here is to get the students to relate the pattern they observe to other interference patterns. Using colored cellophane or a similar filter over a light bulb to approximate monochromatic light provides a simpler pattern for analysis and makes this comparison easier.

STUDENT REFERENCES

Physical Science Study Committee, *Physics*, 2d ed. (Lexington, Mass.: D. C. Heath & Company, 1965), Chapters 17 and 18. Chapter 17 is a concise treatment of interference, emphasizing the simple situation of two point sources. Chapter 18 is a more thorough analysis of interference of light. Several excellent photographs are included.

White, Harvey E., *Modern College Physics*, 5th ed. (New York: Van Nostrand Reinhold Company, 1966). This is another of Professor White's general texts that can be recommended as straightforward, traditional references that are exceptionally good in the general area of optics. In this edition, Chapter 24 discusses interference in sound waves and Chapter 43 deals with diffraction and interference of light.

TEACHER REFERENCES

Physical Science Study Committee, *Laboratory Guide*, 2d ed. (Lexington, Mass.: D. C. Heath & Company, 1965). Experiments II-11 through II-16 could all be considered for laboratory work or for adaptation to demonstrations. Experiments 11 and 12 involve interference of water waves in a ripple tank. Experiments 13 through 16 involve interference of light waves, including double slit and single slit, and applications of interference to resolving power and small distance measurement.

FILMS

Interference of Waves—Ealing Corporation (film-loop). Interference between two point sources of waves is demonstrated with a ripple tank. Source separation and wavelength are identified and varied independently to illustrate their effect on the interference pattern.

Single Slit Diffraction (film-loop).

Multiple Slit Diffraction (film-loop).

Effect of Phase Differences Between Sources (film-loop).

(These film loops from the Ealing Corporation can be used to supplement the first film loop, *Interference of Waves*.)

Diffraction: Double Slit—Ealing Corporation (film-loop). This film loop illustrates interference with light waves and illustrates the effect of varying wavelength and changing the separation between the two slits.

Tacoma Narrows Bridge Collapse—Ealing Corporation (film-loop). This film shows the increasing amplitude of the vibrations in the bridge across the Tacoma Narrows that were caused by inadequate suspension design. That design permitted these vibrations to develop as a result of a forty mile/hr wind. This is a striking example of resonance and is extremely interesting (and fun) to watch, even though the thorough analysis of constructive interference in the torsional mode of vibration in the bridge is complicated. Consequently this film is more suited to mind capture and/or qualitative analysis than to quantitative studies.

Electrostatic Interactions

OBJECTIVE

The student shall be able to predict whether the interaction between bodies with known charges will be attraction or repulsion, and identify unknown charges from their observed interactions with known charges. The student shall also be able to describe and explain the interaction between charged objects and the induced charges upon neutral objects.

DEMONSTRATION

A. Interaction between Charged Objects. Make a small loop out of a strip of paper that is about 2 cm by 6 cm. Suspend the loop from a string or thread that is between 20 and 30 cm long so that a rod can be suspended in the loop and can swing freely. By charging the rod and then bringing other charged objects near it, the attraction between opposite charges and the repulsion between like charges is readily visible. The traditional charging rods, hard rubber rubbed with fur and glass rubbed with silk, work well and can be related to the early definitions of negative and positive charge. Keeping a couple of paper towels between your hand and the silk provides additional insulation, prevents getting the silk sweaty, and makes it easier to charge the glass rod. Plastic strips (Holt, Rinehart and Winston Project Physics materials 03-075005-9, Macalester 1000, or Ealing A27-2906) and their respective charging cloths also work well as known or unknown charged objects.

B. Charging Conductors by Induction. Suspend a charged rod from a paper loop as described in part A. Use this rod to charge an insulated, metal object by induction. By demonstrating the subsequent interaction between the original charge and the induced charge, it will be evident that the induced charge is opposite to the original charge.

Charging an electroscope by induction provides more information about the motion of the charge during this process. The soda straw electroscopes (Macalester 902) are durable, inexpensive, and can be used for demonstration purposes. However, a good quality sensitive electroscope (Ealing A27-3623 or Klinger Ke5235) is an excellent investment for demonstration purposes.

C. Interaction between Neutral Objects and Charged Objects. An interesting follow-up demonstration is to investigate the interaction between a charged rod and the induced charge on an insulated pith ball. The analysis of this situation requires a knowledge of induced charge distributions and the inverse square law. The opposite charge is located on the near side of the pith ball and the like charge will be on the far side of the pith ball. Since the force between charges varies inversely with the square of the distance between them, the force of attraction exceeds the force of repulsion and the pith ball is drawn toward the charged rod.

STUDENT REFERENCES
Moore, A. D., *Electrostatics* (Garden City, N.Y.: Doubleday Anchor Books, 1968). This excellent Science Study Series paperback is a good reference for the student who is not satisfied by the brief treatment of electrostatics that appears in introductory textbooks. It is interesting, and relatively easy reading. Construction details for Moore's "Dirod" electrostatic generators are included for the amateur experimenter or science fair contestant. This book is also recommended for teacher use, both for background information and as a source of demonstration ideas.

TEACHER REFERENCES
Holton, Gerald, and Roller, Duane, *Foundations of Modern Physical Science* (Reading, Mass.: Addison-Wesley Publishing Co., Inc., 1958), Chapter 26. A good historical discussion of electrostatics.
Rutherford, F. J., G. Holton, and F. Watson, Directors, *Project Physics Teacher Resource Book* Unit 4 (New York: Holt, Rinehart and Winston, Inc., 1970). Demonstrations D47 and D48 provide basic demonstrations of electrostatics and charging by induction.

FILMS
Electrostatics (A series of ten film loops)—Ealing Corporation. These films could be used in class for demonstrations, for self-study, or for testing purposes in some cases.
Coulomb's Law—Modern Learning Aids (30 minutes). Professor Eric Rogers demonstrates the inverse square law and explains an inverse-square relationship with his remarkably clear "butter-gun" analogy. The test of the inverse square law based on searching for charges inside a charged hollow sphere is also shown. This is a PSSC film.

Lines of Force of Electrostatic and Magnetic Fields

OBJECTIVE
The student shall be able to sketch some lines of force to represent the electric field between two simple electrodes and some lines of force to represent the magnetic field around a magnet and a current-carrying wire.

DEMONSTRATIONS
The overhead projector can be used to introduce students to the concept of lines of force for both electrostatic and magnetic fields.

A transparent compass works well on the overhead projector except that the steel parts of the projector will usually prevent the compass from aligning exactly in the earth's field. Iron filings can also be used to illustrate the field pattern around magnets. Use heavy tape to mark one end of the magnets so that their polarity can be identified by the students. The magnetic field around a wire or a solenoid (Ealing A27-0322) can also be demonstrated, by using a nickel-cadmium or auto battery to provide the large current required.

For the electrostatic field demonstration, obtain a clear glass or plastic tray that is reasonably free of distortion. Pour clear oil into the tray until the oil is about one-half inch deep. Use vacuum pump oil, mineral oil, castor oil, or any other clear oil. Gently stir in about 2 ml of grass seed, fine hair, or fiber clippings. Try your demonstration and add more seeds or fibers if needed.

The electrostatic field can be provided by a Wimhurst, Van de Graaff, or Dirod machine. Form electrodes of various shapes from lightweight sheet metal. The field pattern between point electrodes, parallel plates, inside and outside of a closed conductor, and near a point can all be demonstrated. The electrostatic field causes the seeds or fibers to become electric dipoles that become aligned in the electrostatic field in a manner analogous to that in which iron filings become aligned in a magnetic field.

The action of a lightning rod can be demonstrated using a Wimhurst or Dirod machine. A dressmaker's pin will stand upright if it is pushed through a small cork. The point of the pin should extend about 1 cm so the students can see it. Find a *smooth*, metallic object that is approximately the same height as the pin and place it on an insulated metal plate. Clamp another metal plate above the first so that there is about 1 cm between it and the metallic object. (Welch #1968 can be used for the metal plates, or use the metal plates with insulated handles from two electrophorouses.) Check to be certain that the students can see the metallic object. Connect the Wimhurst, Van de Graaff, or Dirod machine to the plates and operate until large sparks jump between the metallic object and the upper plate. Stop just before a spark is formed and discharge the plates. The discharging process should provide a spark.

Place the pin, held upright by a cork, beside the metallic object on the lower plate and repeat the charging and discharging process. This time there should be no spark because charge is transferred between the plates via ions that are formed in the strong field near the point of the pin. The fact that no spark is produced when the plates are discharged should convince your students that the charge has been transferred between the two plates.

STUDENT REFERENCES

Dart, Francis E., *Electricity and Magnetic Fields* (Columbus, Ohio: Charles E. Merrill Publishers, 1966). This paperback is part of Merrill's paperback physical science series aimed at the college nonscience major. It is readable, includes problems, and uses only algebra for mathematics.

Physical Science Study Committee, *Physics*, 2d ed. (Lexington, Mass.: D. C. Heath & Company, 1965), Chapters 27 and 30. Electric field patterns are extensively discussed. Figure 27-9 is a photograph of grass seeds in oil with different applied electric fields.

Rutherford, F. J., G. Holton, and F. Watson, Directors, *The Project Physics Course Text* (New York: Holt, Rinehart and Winston, Inc., 1970), Chapter 14. A concise treatment of electrostatic and magnetic fields.

TEACHER REFERENCES

Rutherford, F. J., G. Holton, and F. Watson, Directors, *Project Physics Teacher Resource Book* Unit 4 (New York: Holt, Rinehart and Winston, Inc., 1970). Another field mapping technique, using a 45-volt battery as a potential source and using two probes connected to a microammeter to detect equipotential lines in the region between electrodes in a pan of water, is described.

Feynman, Richard P., Robert B. Leighton, and Matthew Sands, *The Feynman Lectures on Physics* (Reading, Mass.: Addison-Wesley Publishing Company, Inc., 1964) Volume II, Chapter 9. This chapter discusses atmospheric electricity and lightning.

Effect of a Magnetic Field Upon Moving Charged Particles.

OBJECTIVE

The student shall be able to indicate the direction of the force, if any, that is exerted upon particles of known charge, given their direction of motion and the direction of the applied magnetic field.

DEMONSTRATION

Using a Tesla coil (CENCO No. 80730) as a high voltage source, the path of an electron beam is readily visible in a Crookes tube with a fluorescent screen (CENCO No. 71555). The direction of the motion of the negative particles is from the high voltage electrode toward the grounded electrode. A good bar magnet can be used as the source of the applied magnetic field, which is away from the North magnetic pole and toward the South magnetic pole. The deflection of the electron beam in the vertical plane is readily observed as the bar magnet is held in the horizontal plane, perpendicular to the beam. These observations can be used as the basis for the hand rule that is used in your text.

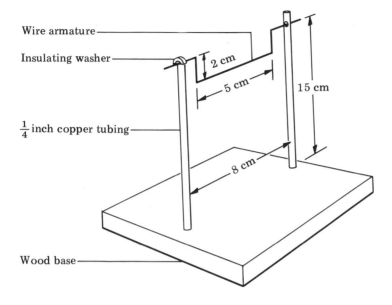

Figure 3-11. Construction details of a "Bergsten motor."

A good follow-up to the introductory demonstration is a simple "Bergsten motor,"* which is constructed of two pieces of copper tubing about 15 cm and 17 cm long, respectively. A small hole, about 1/8 inch, is drilled through the longer tube 2 cm from one end. Glue one half of an insulating washer to the top of the shorter tube to serve as a "commutator." Mount an "armature" made of a piece of stiff uninsulated wire as shown in Figure 3-11.

Use a battery capable of delivering a large current because the circuit has very little resistance. Hold one pole of a strong magnet (CENCO No. 78329) beneath the wire and close the circuit. One piece of tubing is connected to the positive terminal and the other piece of tubing is connected to the negative terminal. The effect of the magnetic field on the current in the "armature" wire should force it sideways. If the wire is given a push to start it, it should continue to move in a circle. As the wire moves upward, it will rub against the fiber washer and the circuit will be broken. The wire will then continue its motion over and downward until it comes into contact with the tubing again, and the cycle will repeat itself. A little practice is required to find the most effective position for the magnet.

---

*This demonstration was first shown to us by Ronald Bergsten of the University of Wisconsin—Whitewater physics department.

A similar demonstration uses an apparatus that has two metal rods (aluminum, brass, or ¼ inch copper tubing) about 30 cm long. The rods are placed parallel to each other and are spaced by about 10 cm. One of the rods is connected to the positive terminal and the other to the negative terminal of a battery or a similar voltage source that is capable of providing a large current. If a short metal rod is laid perpendicularly across these two rods, a large current flows through it. By applying a strong magnetic field in the vertical direction, the short rod can be "pushed" along the parallel tracks. The students should be able to predict the direction of this motion by using the hand rule given in their text.

STUDENT REFERENCES

Rutherford, F. J., G. Holton, and F. Watson, Directors, *The Project Physics Course Text* (New York: Holt, Rinehart and Winston, Inc., 1970), Chapter 14 and 15. Chapter 14 presents one version of the hand rule for describing the interaction between a magnetic field and moving charges. Chapter 15 includes a discussion of the operation of motors and generators.

Physical Science Study Committee, *Physics*, 2d ed. (Lexington, Mass.: D. C. Heath & Company, 1965), Chapter 30. A hand rule is presented. Several of the diagrammed examples are similar to the simple demonstrations presented here. Meters, motors, and mass spectrometry are discussed as illustrations.

TEACHER REFERENCES

Trout, Virdine E., "The Motor Effect," *The Science Teacher* 35 (6):74-76 (Sept., 1968). This apparatus note is a discussion of two demonstrations that are very similar to the last demonstration described here.

Rutherford, F. J., G. Holton, and F. Watson, Directors, *Project Physics Teacher Resource Book* Unit 4 (New York: Holt, Rinehart and Winston, Inc., 1970). Demonstration D50 uses the Project Physics current balance to illustrate interactions between applied magnetic fields and the current in the wire of the apparatus. Experiments E35, E36, and E37 all deal with the interaction between currents and magnetic fields; E37 is unique in that the students first construct their own electron beam tube (Holt, Rinehart and Winston 03-074980-8) and then observe electrostatic and magnetic deflection procedures.

FILMS

Electrons in a Uniform Magnetic Field—Modern Learning Aids (11 minutes). In this PSSC film, the path of electrons in a cathode ray tube is shown and their deflection in an applied magnetic field is illustrated. With reference to the charge value from the Millikan experiment, a value for the mass of the electron is computed from the collected data.

Synchrotron (14.5 minutes). Holt, Rinehart and Winston, Inc. (It is also available free from the U.S. Atomic Energy Commission Film Libraries but no notes are included with the film.) The components and operation of the Cambridge electron accelerator are shown and explained in this Project Physics film.

Radioactive Decay

OBJECTIVE

Given some data from a radioactive sample, the student shall be able to plot a counting rate versus time curve, and use this curve to predict the counting rate at some time beyond the interval during which the data was collected.

DEMONSTRATION

Some short-lived radioactive isotopes occur naturally. To collect them, cover the air intake of a vacuum cleaner with a piece of wire screen and secure a paper or cloth filter over this screen. The best place for the vacuum cleaner is in a basement or closet where the ventilation is poor. Run the machine for 2-4 hours before the demonstration. The dust that is collected on the filter will contain decay products of radon and thoron. Handle the filter carefully so that the dust is not dislodged.

At the time of the demonstration, clamp a Geiger-Mueller tube to a ring stand, record the background count over several minutes, and determine the average background counting rate. Use the most sensitive equipment available since the activity of the source will be very low. Place the filter close under the Geiger-Mueller tube and count for 1-minute intervals.

The primary activity of the dust on the filter is usually from ^{214}Pb and ^{214}Bi, which are decay products of radium having half-lives of 26.8 minutes and 19.7 minutes, respectively. Both are beta emitters. Because several radioactive materials are present, the decay curve is not a simple decay curve. The first data taken, however, does yield a fairly good decay curve with an effective half-life of 45-50 minutes. (See McGinley). This method thus provides a convenient method for obtaining a radioactive sample whose decay can be observed in one class period, and a basis for a discussion of the sources of background radiation.

STUDENT REFERENCES

Hermies, Sister Mary, and Sister Mary Joecile, *Radioactivity: Fundamentals and Experiments* (New York: Holt, Rinehart and Winston, Inc., 1963). This paperback book contains descriptions of measuring techniques and instructions for performing many experiments.

TEACHER REFERENCES

McGinley, Patton H., "Half-Life of Dust" *The Physics Teacher*, Sept. 1968, p. 323. The collection of dust by air filtration is described. The half-life reported for their dust sample is 45 minutes.

Robert S. Shankland, *Atomic and Nuclear Physics* (New York: The Macmillan Company, 1960), Chapters 8 and 12. These two chapters contain a discussion of radioactive series and of cosmic rays, the other major source of background radiation.

FILMS

Half-life of a Radioactive Coin—Ealing Corporation (film-loop). A silver disk is activated by a neutron source, and the disk's radioactivity is measured as it decays. This is a Project Physics film.

^{238}U *Radioactive Series*—McGraw-Hill (9 minutes). This film traces the various stages in the decay of ^{238}U to stable lead.

Contemporary Physics

OBJECTIVE

The student shall be able to recognize that active research at the frontiers of knowledge is a very important part of physics, and that the physicists participating in these activities are human beings as well as scientists.

DEMONSTRATION

Unless you can arrange a tour of a university or an industrial research laboratory with a scientist as a tour guide, some of the following films are probably the best means of achieving this objective. This topic is also an appropriate concluding note to a physics course.

REFERENCES

Gamow, George, *Thirty Years That Shook Physics* (Garden City, N.Y.: Doubleday-Anchor Books, 1966). This is a description of the development of quantum mechanics by a man who was an active participant in this process. Along with a case-study history of this development, Gamow includes sketches and personal anecdotes that provide considerable insight into this period and the men who shaped this new discipline.

Fermi, Laura, *Atoms in the Family* (Chicago: University of Chicago Press, Phoenix Books, 1954). This is the biography of one of the greatest physicists of all time, Enrico Fermi, as written by his wife. The dedication and activities of the scientist are clearly described in this book, which should be of interest to both science- and nonscience-oriented students.

FILMS

The World of Enrico Fermi—Holt, Rinehart and Winston, Inc. (47 minutes). This documentary film, which describes Fermi's life and work, was made for Project Physics from old documentary films and still photographs. The viewer meets almost all of the great physicists of this era, and their relationship to Fermi and his work is discussed. This is an excellent film, which deserves a wider audience than just physics classes—who should *certainly* see it.

People and Particles (28 minutes). Holt, Rinehart and Winston, Inc. (It is also available free from the U.S. Atomic Energy Commission Film Libraries.) The activities and members of a particle physics group are shown. An effort is made to provide insight into the professional and personal lives of these scientists, both the faculty and the graduate students. This is a Project Physics film.

Laser Light—Scientific American (37.5 minutes). This is a beautiful film. Laser demonstrations and their applications in technology are shown. This film could be used to emphasize the rapid transition of developments such as the laser from scientific to technological interest.

PART...4

Additional Resources for the Teacher of Chemistry and Physics

Performance Objectives

Performance objectives have been on the educational scene for about ten years. These objectives are a direct outgrowth of the programmed instruction movement that was begun in the 1950's by B. F. Skinner. Performance objectives provide the teacher with an excellent tool for quantizing the content and/or skills that are to be learned by the student. The teacher will know exactly what content will be taught and also what cognitive level of response he wants from his students. (Bloom, 1956) If a list of the content and skill objectives is also shared with the student, the student knows what the teacher feels is important for him to learn. The student will also be aware of what cognitive level of performance will be expected of him by the teacher. The teacher and student can then work together to improve the efficiency of the teaching-learning process.

When the teacher prepares his own performance objectives, he is required to analyze the course of instruction. An analysis of a chemistry or physics course of study will produce the following: (a) the major units will be divided into topics, (b) the topics will be divided into principles, and (c) the principles will be divided into concepts. Specific performance objectives for each level will yield a hierarchy that will direct the content of each day's lesson. An examination of the initial objectives may reveal that some of them are trivial or that some need to be combined or rewritten to provide an objective that has a viable rationale for being included in the course of study. Many teachers avoid writing performance objectives in the hope that they will escape an evaluation of what they are teaching. Other teachers see it as a means of organizing instructional sequences into a meaningful pattern to increase the learning rate of the student.

When the teacher knows the sequence of instruction and the performance objectives that are to be attained by the students, the mode of instruction can be selected that will maximize the probability of success by those students. After the students have been tested as to their actual acquisition of those skills, the instructor can then evaluate both the performance objective and the instruction. The performance objective can be evaluated with respect to its applicability to the course and its position in the instructional sequence with respect to other objectives. Clarity of the performance objective to the students should also be considered. If the objective is found to be valid, then the instruction can be evaluated with regard to student success on tests constructed from those objectives. The teacher may decide that the particular mode of presentation was appropriate or that a new approach is required. Tentative proposals should be included with each performance objective if revision is necessary before future use.

The students profit by having a list of performance objectives. This list will be used to guide their study of the material by providing them a reference for relating an individual objective to the total instructional sequence. Students may also assist in providing constructive criticism as to the appropriateness of the objectives as the concepts are presented. For example, the student may already be capable of performing at a specified level and hence would not need to be instructed in that area again. The performance objectives should increase student learning efficiency by focusing instruction on problem-solving instead of the dispensing of information. If the teacher considers a topic to be important, the student will probably agree.

Writing performance objectives requires time and practice. It cannot be done without much consideration as to what is really important for the student to learn. As the reader considers the following materials, he is urged to examine the concept of performance objectives in detail and to relate the writing of performance objectives to the improvement of his instruction.

These materials were designed to help the teacher to learn how to write specific performance (behavioral) objectives. If you can already write three-parted objectives that are consistent with the four rules given later in this section, turn to Part F of this exercise and begin to write. These materials will provide you with the following practice:

A. Identifying elements of the performance objectives.
B. Identifying missing and incorrect elements.
C. Listing cognitive, affective, and psychomotor verbs.
D. Writing test situations.
E. Qualifying objectives.
F. Writing specific performance objectives congruent with general ideas.

Proceed through the materials at your own pace. If you prefer working with a partner, *please do*!

Objective: Given a general description of some desirable student performance, you should be able to write performance objectives for the performance that contains a testing situation, a measurable performance term, and qualifying terms.

Definition: Performance objectives are statements written in measurable terms that describe precisely what a student should be able to do after he receives the instruction that you are planning.

The specific nature of performance objectives is described by the following four rules:

Rule 1. Performance objectives should contain at least three elements: (a) a description of the testing situation, (b) a measurable performance term, and (c) qualifying terms stating the level of acceptable performance.

Rule 2. The testing situation should state explicitly what the student will be given when his performance is being measured. It describes information, materials, equipment, opportunities, situations, and the like. It must be congruent with the performance term and qualifiers.

Rule 3. The performance term is a measurable statement such as construct, name, compare, list, recite, or apply. The word *measurable* is meant to imply that the words used have a common definition—that is, the words would be defined in the same way by almost everyone. It describes the actions or behaviors that the student is to exhibit.

Rule 4. The qualifying terms describe the level of acceptable student performance. They should explicitly delimit the stated performance—spell out the level of sophistication, degree of accuracy, or level of affectivity. They must be congruent with the testing situation and qualifiers. They should define acceptable performance precisely without providing answers that students could memorize and thereby satisfy the requirements of the objective.

A. Examine this objective: (Given a general description of some desirable student performance), you should be able to *write* performance objectives for the performance [which contain a testing situation, a measurable performance term, and qualifying terms.] Does the objective:

a. contain a testing situation, performance term, and qualifying terms? (Yes, No)
b. describe what the student will be doing when his performance is being measured? (testing situation) (Yes, No)
c. contain a measurable performance term? (Yes, No)
d. explicitly state the level of performance sophistication without providing a memorizable answer? (qualifying term) (Yes, No)

The sample objective is appropriately stated. You should have answered yes to all the questions. If you did not, please read the rules again and reexamine the objective.

Subobjective One: Given three objectives, you should be able to identify the testing situation, performance term, and qualifying terms. Acceptable performance would include correctly identifying eight of nine elements in the three objectives.

Read the three following objectives. Place parentheses ( ) around the testing situation, *underline* the performance term, and place brackets [ ] around the qualifying terms.

1. Given ten performance objectives, you should be able to label the testing situation, the performance term, and the qualifying terms. Acceptable performance would include correctly labeling twenty-seven of the thirty elements in the ten objectives.
2. Given a slide, coverslip, and a culture of bacteria, the student should be able to prepare a hanging-drop slide. Acceptable performance would include making a drop that will not fall off.
3. Given a question concerning the factors that affect fossilization, the student should be able to state or list five factors that may affect fossilization.

Check your answers with the answer response sheet on page 106. Do not refer to the answer sheet until you have completed the work.

B. Read the objectives on the next few pages and respond to the questions asked, using sheets of ruled paper. Complete all of the first Practice Set before looking at the answer sheet.

Subobjective Two: Given objectives, you should be able to discriminate between objectives that contain all elements and those that do not identify the missing or incorrect element. Acceptable performance would include correctly discriminating between all complete and incomplete objectives and identifying 90 percent of the missing or incorrect elements.

Practice Set 1

Objective 1. The student should be able to write performance objectives.
_____a. This is an acceptable performance objective
_____b. This is not an acceptable performance objective because

---

Objective 2. Given several performance objectives, the student should be able to identify the testing situation, performance term, and the qualifying terms in each.
_____a. This is an acceptable performance objective
_____b. This is not acceptable because_____

Objective 3. Given access to a large variety of books, the student should demonstrate an appreciation of science by reading a book on science in his free time. (Acceptable performance would include reading one book each semester.)
    _____a. This is an acceptable performance objective
    _____b. This is not acceptable because_____

Objective 4. Given a slide, cover slip, and culture of bacteria, the student should be able to prepare a hanging-drop slide preparation.
    _____a. This is an acceptable performance objective
    _____b. This is not an acceptable objective because_____

Objective 5. Given a pencil and paper, the student should be able to define the term *atom* in his own words.
    _____a. Acceptable performance objective
    _____b. Unacceptable because_____

Objective 6. Given several unidentified insects and a taxanomic key, the student should be able to key out 80 percent of the insects to the correct order.
    _____a. Acceptable performance objective
    _____b. Unacceptable because_____

Objective 7. You should be able to determine the value of "g." Acceptable performance would include obtaining an answer within 5 percent of the value of "g" stated in your text.
    _____a. Acceptable performance objective
    _____b. Unacceptable because_____

Compare your answers with those on the answer sheet. If you have any questions concerning the answer, see your instructor. Correct your wrong answers. Proceed to Practice Set 2. *Do not* look at any answers until you have completed the entire set.

Practice Set 2

Objective 1. Given an opportunity the student should freely verbalize an appreciation of science.
    _____a. Acceptable performance objective
    _____b. Unacceptable because_____

Objective 2. After completing a unit on the earth, the student should be able to provide evidence supporting the idea that the earth is round. Any two statements of evidence will be acceptable performance.
    _____a. Acceptable performance objective
    _____b. Unacceptable because_____

Objective 3. Given a textbook definition of "spontaneous generation," you should be able to rewrite the definition in your own words.
    _____a. Acceptable performance objective
    _____b. Unacceptable because _____

Objective 4. Given a laboratory, the student should be able to design an experiment to determine if "spontaneous generation" of microorganisms will occur. Acceptable performance would include designing a controlled experiment that could be conducted in the school laboratory.
    _____a. Acceptable performance objective
    _____b. Unacceptable because _____

Objective 5. Given a picture of an actual stream, the student should be able to identify the stage of development of the stream and support his conclusion. Acceptable performance would include correctly identifying the developmental stage of the stream and listing three identifying features that support his conclusions.
    _____a. Acceptable performance objective
    _____b. Unacceptable because _____

Objective 6. Given a slide (35mm) of a partially split rock located in a desert climate, the student should be able to describe how the splitting took place.
    _____a. Acceptable performance objective
    _____b. Unacceptable because _____

Objective 7. Given a question concerning the factors that affect fossilization, the student should be able to state or list five factors that may affect fossilization.
    _____a. Acceptable performance objective
    _____b. Unacceptable because _____

Check your answers. If you did not miss any of the questions, proceed to the next part. If you missed any of the questions, start from the beginning.

C. Prepare lists of appropriate verbs (Rule 3). The heart of the performance objective is the performance term—the definition of the specific action or behavior that the student should be able to perform. These words must generally have *simple operational definitions*—definitions that are understood by everyone. If the student does not understand what you want, the likelihood that he will perform at the desired level is greatly reduced. Hence, it is essential to use only words that have simple, commonly agreed upon, operational definitions. Prepare lists of measurable action verbs that may be used to define Cognitive, Affective, and Psychomotor behaviors.

Subobjective Three: When asked, you should be able to provide examples of Cognitive, Affective, and Psychomotor verbs. Acceptable performance includes providing at least ten of each variety.

Cognitive Verbs	Affective Verbs	Psychomotor Verbs
1.	1.	1.
2.	2.	2.
3.	3.	3.
4.	4.	4.
5.	5.	5.
6.	6.	6.
7.	7.	7.
8.	8.	8.
9.	9.	9.
10.	10.	10.

a. Compare your list with the lists constructed by other members of the class. Add words to your list.
b. Compare and contrast your list with the list on the response sheet.

D. Describe (write) testing situations (Rule 2). Rule 2 states that the testing situation describes what the student will be doing when his performance is being measured. It may describe the physical environment, the materials or information that you will provide the student when his performance is being measured.

Subobjective Four: Given the performance terms of an objective, you should be able to write a description of what you might provide the student and what he will be doing when his performance is being measured. Acceptable performance includes preparing statements consistent with the definition stated in Rule 2.

1. $F = ma$ (The student should be able to calculate the value of A.)
2. The student should be able to calculate the density of a uniform cube of a substance.

3. The student should be able to calculate the molecular weights of the substances.
4. The student should read articles and books about science and scientists.
5. The student should be able to design a plan for monitoring the oxygen content of a pond.
6. The student should be able to construct hypotheses that could be tested and might identify the causes of the pollution.
7. The student should be able to paraphrase "ontogeny recapitulates phylogeny."
8. The student should be able to label the four chambers of the heart.
9. The student should be able to identify the five stages of miosis.
10. The student should participate in some extracurricular science-related activity.

Reexamine the testing situations that you have prepared. Do they all describe what the student will be doing when his performance is being measured? Compare your situations with the situations provided with Part E. Since it may be possible to have several different testing situations, your statements may not be the same as those stated.

E. Qualify performance statements, Rule 2.

Qualifying Terms: Commonly used action verbs may have different meanings in different contexts. Examine this simple example: *"Define the word atom."*

What might this request mean—

to an eighth-grade student?
to a chemistry student?
to a graduate student in chemistry?
to an atomic physicist?
to a teacher?

The word *define* is a measurable performance term. However, it does not provide enough guidance; it must be qualified.

Writing the qualifier is often the most difficult task because the qualifier must be written in such a way that it provides explicit guidance without providing an answer that could simply be memorized. For example:

1. When asked, the student should be able to define the word *atom*. Acceptable performance would include defining an atom as the

chemical unit of matter that is the smallest particle of an element able to enter into a chemical reaction.

To fulfill this performance, all the student must do is memorize and prepare to regurgitate your objective.

Be subversive. Tell your students exactly what to do and pat them on the back when they do it. *But* do you have as an objective *"memorize my objectives?"* Examine the following way to state the same objective:

2. When asked, the student should be able to define the word *atom*. Acceptable performance includes citing the type of unit an atom is, its size, and the relationship between atoms and chemical reactions.

The second objective tells the student what he is to do without permitting him to memorize your objective.

Subobjective Five: Given a testing situation and the performance terms, the student should be able to write qualifying terms. Acceptable performance would include preparing statements that would explicitly define the level of performance expected without providing a memorizable answer.

1. Given the formula $F = ma$ and the values of F and m, the student should be able to calculate the value of a.
2. Given a single pan balance and a ruler, the student should be able to calculate the density of a uniform cube of a substance.
3. Given a table of atomic weights and the chemical formulas of compounds, the student should be able to calculate the molecular weights of the compounds.
4. Given access to a library, the student should read books and articles about science and scientists.
5. Given description of a pond and a list of available equipment and supplies, the student should be able to design a plan for monitoring the oxygen content of a pond.
6. Given photographs of a polluted stream and descriptions of the area through which the stream passes, the student should be able to construct hypotheses that could be tested that might identify the causes of the pollution.
7. When asked, the student should be able to paraphrase the phrase, "Ontogeny recapitulates phylogeny."
8. Given a diagram of the heart, the student should be able to label the four chambers.

9. Given twenty photographs of cells in different stages of mitosis, the student should be able to identify examples of the five stages of mitosis.
10. Given free periods of time during the school year, the student should participate in some extracurricular science-related activity.

Different teachers will usually qualify objectives in different ways. Reexamine the definition of the qualifying terms again, and then compare your qualifiers with those given on the answer sheet.

F. Ideas to objectives.

Terminal Objective: Given a general description of some desirable student performance, you should be able to write performance objectives for the performance that contains a testing situation, performance term, and qualifying terms.

Write at least two objectives for each general statement.

General Statement No. 1: A teacher wants his students to understand Newton's Third Law of Motion, which Newton stated as follows: "To every action there is always opposed an equal reaction; or, the mutual actions of two bodies upon each other are always equal and directed to contrary parts."

General Statement No. 2: A teacher wants his students to know the water cycle.

General Statement No. 3: A teacher wants his students to demonstrate an appreciation of science.

General Statement No. 4: A teacher wants his students to have a feeling for geologic time.

General Statement No. 5: A teacher wants his students to intelligently discuss atoms, molecules, and the relationships between atoms and molecules.

General Statement No. 6: The teacher wants to plan a field trip to the local sewage treatment plant.

General Statement No. 7: The teacher wants his students to truly understand man's environmental problems.

General Statement No. 8: The teacher wants his students to plan a Chemistry experiment they could conduct.

General Statement No. 9: A teacher wants his students to know about neutralization reactions. (Salt + Water, Acid + Base, etc.)

106 Additional Resources for the Teacher of Chemistry and Physics

General Statement No. 10: A teacher wants his students to participate in extracurricular science-related activites.

Have a fellow student or teacher examine and criticize all your objectives.

Answer Sheet

PART A
1. (Given ten performance objectives), you should be able to <u>label</u> the testing situation, the performance term and the qualifying terms. [Acceptable performance would include correctly labeling twenty-seven of the thirty elements of the ten objectives.]
2. (Given a slide, coverclip, and a culture of bacteria), the student should be able to <u>prepare</u> a hanging-drop slide. [Acceptable performance would include making a drop that will not fall off.]
3. (Given a question concerning factors that affect fossilization), the student should be able to <u>state</u> or list [five] factors that may affect fossilization.

PART B (PRACTICE SET 1)
1. <u>b</u> (See Rules 1, 2, and 4. The testing situation and qualifier are absent.)
2. <u>a</u> (Each qualifies and means all. If you marked (b) you should have indicated a need for a better qualifier.
3. <u>a</u> (Acceptable)
4. <u>b</u> (A qualifier is needed to define an acceptable preparation.)
5. <u>b</u> (A qualifier is needed.)
6. <u>a</u> (Acceptable; 80 per cent is the qualifier.)
7. <u>b</u> (The testing situation is not specified.)

PART B (PRACTICE SET 2)
1. <u>b</u> (Qualifier needed; "appreciation" could be negative or positive.)
2. <u>b</u> (Testing situation not given.)
3. <u>a</u> (However, the qualifier, "in your own words," is not too good and "b" is probably also an appropriate response.)
4. <u>b</u> (Testing situation not appropriate; a laboratory is not absolutely necessary for designing.)
5. <u>a</u> (Acceptable)
6. <u>b</u> (Needs qualifier)
7. <u>a</u> (Acceptable; however, you may question if simply adding the five is adequate qualification)

PART C
Measurable Performance Terms:

Cognitive

recall	describe	analyze
define	compare	interpret
state	contrast	extrapolate
recite	design	translate
identify	evaluate	synthesize
name	apply	design
hypothesize	calculate	list

Affective

receive	participate	values
attend	joins	responds
select	delay	commitment
uses	supports	seeks

Psychomotor

constructs	draws	sets up
manipulates	makes	prepares
measures	transfers	diagram
weighs	uses	

PART E
1. One-half credit will be given for setting up the problem and half credit will be given for the correct answer.
2. Acceptable performance would include obtaining a value within 5 per cent of the value given in the Handbook of Chemistry and Physics.
3. Acceptable performance would include determining the correct molecular weight of eight of ten of the compounds.
4. Acceptable performance would include reading one book or three articles each semester.
5. Any plan that would provide a quadradaily record accurate to within 5 per cent for a period of six months would be acceptable.
6. Acceptable performance includes constructing three testable hypotheses.
7. Acceptable performance includes defining each word in language common to the secondary school student.

8. Acceptable performance includes labeling the right and left atria and the right and left ventricles.
9. Acceptable performance would include correctly labeling sixteen of the twenty photographs.
10. Acceptable performance would include participating in any single science-related extracurricular activity each semester.

ADDITIONAL REFERENCES

Armstrong, Robert J., Terry D. Kramer, and E. Wayne Robertson, *Developing and Writing Behavioral Objectives* (Tucson, Aris.: Educational Innovators Press, Inc., 1968).

Ebenson, Thorwald, "Writing Instructional Objectives," *Phi Delta Kappan* (January, 1968).

Eiss, Albert F., and Mary Blatt Harbeck, *Behavioral Objectives in the Affective Domain* (Washington, D.C.: National Science Teachers Association, 1969).

Gronlund, Norman E., *Stating Behavioral Objectives for Classroom Instruction* (New York: The Macmillan Company, 1970).

Kurtz, Edwin B., "Help Stamp Out Non-Behavioral Objectives," *The Science Teacher* (March, 1968).

Mager, Robert F., *Preparing Instructional Objectives* (Palo Alto, Calif.: Fearon Publishers, Inc., 1962).

Popham, W. James, and Eva L. Backer, *Establishing Instructional Goals* (Englewood Cliffs, N.J.: Prentice-Hall, Inc., 1970).

Popham, W. James, et al., *Instructional Objectives*, American Educational Research Association Monograph (Chicago: Rand McNally and Company, 1969).

Tanner, Daniel, *Using Behavioral Objectives in the Classroom* (New York: The Macmillan Company, 1972).

Lesson Planning

A successful teacher plans his students' progress from the beginning of each class session to the end. Instruction is not a matter of luck or a function of the amount of time spent in activity; it requires careful planning and preparation. Efficiency or maximum utilization of the available time should be the objective of the successful teacher.

Planning for a chemistry or physics class must involve a greater effort by the teacher than merely a brief review of the chapter before beginning the daily class session. Each day's instruction must be part of a unit of instruction that is based upon the year's objectives.

Written lesson plans demand a considerable amount of time and many teachers dislike spending time writing out formalized plans. These teachers prefer instead to scribble a few notes on a piece of paper before class time or to circle key problems that will demonstrate basic principles to work out in class. Such a teacher will typically lecture his class rather than actively involve the students in a discussion or problem-solving session. This teacher rarely considers the motivating forces that are produced by the utilization of different teaching strategies that can improve student understanding.

Why should the novice or experienced teacher spend the time required to write a "good" lesson plan?

1. The teacher can see a broad picture of the material to be studied over a given time period.
2. The teacher may state the course objectives in measurable terms.
3. It will provide the teacher with the opportunity to assess the prerequisite skills of the students.
4. Supplementary materials can be secured.
5. References may be checked.
6. Realistic assignments may be planned.
7. Different teaching strategies can be implemented.

The lesson plan for a given day's instruction is a product of a series of selections by the teacher. The teacher must first assess the prerequisite capabilities of each student, either directly or indirectly, to determine where the student can successfully enter the structure of the course. There exists, either on paper or in the teacher's mind, a broad outline of the topics to be learned by the student during the year. These topics may be sequenced into a logical order that should reflect the most efficient order for learning. Ideally, the first topic of instruction will be one that has commonalities with previously learned concepts. Through this review and the addition of new material, the student will gain the confidence required to successfully assimilate the new concepts. Since a topic will require more than one day in the classroom, the teacher must further divide the topic into smaller units that are logically ordered to insure student success. The daily lesson plan consists of the appropriate subdivisions of the topic.

Sequencing will result from either the teacher's knowledge of the logical structure of the material or by following some other plan of organization. One such plan that is frequently followed is the order of the textbook. However, most textbooks contain many more topics than students can learn well in one year. The teacher may cover the material in any given textbook but this does not mean that the students have learned the material. The teacher should be prepared to guarantee the student a high probability of success of learning whatever is selected for study. Student success requires a statement of the specific behavior changes that are anticipated from any given day of instruction. When these statements are made, a logical structure usually emerges.

A statement of the specific performance objectives allows the teacher and student to strive for some attainable goal. Instruction moves from just a brief contact between the student and the concept to a specified use. Being able to apply a principle is a different learning task than merely knowing that the principle exists. If the students are to strive for the higher levels of understanding, as outlined by Bloom in *Taxonomy of Educational Objectives*, the teacher must know the types of student behavior changes that will be expected. When this is known and stated, evaluation questions can be constructed that will require the students to express behaviors reflecting this higher level of understanding. The student's success is increased even more when he knows what types of behavior are being expected of him.

The student's success depends in part on his ability to relate new concepts to ones that he already possesses. The teacher must be aware of the prerequisite capabilities required to successfully learn a given concept. If the student does not have these prerequisites, then some provision should be made for him to acquire them. The length of time required for the successful learning of a concept will depend upon the prerequisites. For an advanced class the teacher need give only a brief presentation, whereas more time must be spent with a group that does not have the background.

Lessons prepared in advance allow the teacher time that he needs to prepare adequately. This preparation can include the preparation of solutions or the acquisition of equipment if the school lacks what the teacher feels he needs to adequately present the topic. Supplementary materials may be prepared and available in the classroom. This gives the teacher the flexibility of presentation in the classroom to provide for individual differences.

References are often needed when the students are allowed to question extensively. The new teacher will learn more by pursuing the questions of his students than he did in his college training. An understanding of the interrelationships of concepts is also gained by the teacher. Advanced planning provides time for a review of the basic principles as well as conventions that are forgotten if not used. Reference books in the classroom can provide particular pieces of information during class so that the teacher can respond to questions that were not anticipated. If references are available in the library as well as in the laboratory, the students may be assigned research investigations. A reference book will frequently have a particularly outstanding explanation that should be brought to the students' attention.

Problem assignments are an integral part of the chemistry and physics courses. Processes are learned and concepts are reinforced by problem-solving. The student can either do these problems in a mechanical trial and error method or he can actually understand the process. It is difficult to distinguish the difference unless the teacher has coordinated the problem assignments with the class activities. The teacher must then use those lengthy assignments to generate questions or expose difficulties that were not apparent in the original presentation.

Utilizing different teaching strategies will necessitate considerable expenditure of time by the teacher. Student ability and teaching materials must be matched in the most logical manner. The strategy selected will often require the acquisition of materials that are foreign to a lecture presentation. The increased efficiency of learning by the students justifies the teacher preparation.

Varied strategies such as laboratory, problem solving, case histories and the like will also serve to generate new interest because they are different. New patterns of instruction require an increased mental effort by the student. The student's mind seeks to accommodate the new instructional strategy with the old strategy to determine if the added effort to conform to the instruction will be rewarding. A new strategy must not be used continuously as it quickly loses its attractiveness to the student. The objective is to make the student anticipate the uncertain factor of the class. Several strategies must be used to carry this out. From year to year the teacher can prepare new lesson plans for the same content material using different strategies. This will enable him to select a different mode of instruction in future years depending upon the class and other variables that are not clearly defined. Success of one strategy with one group of students does not imply success of the same strategy with another group.

Several possible lesson plan formats might be used. The following is presented with examples for the teacher's consideration. The basic components are all included but the order may be changed to suit the individual teacher's situation.

Lesson Plan Format

Topic: Date used:
Audience:
Instructional Strategy:
    I. Objectives
       A. Rationale
       B. General Objectives
       C. Specific Performance Objective
   II. Procedures
       A. Approach
       B. Development
       C. Summary Questions
  III. Assignments
  IV. Supplies
       A. Audiovisual
       B. Laboratory
   V. Reference Materials
  VI. Evaluation

The objectives are the framework for the lesson plan as they determine the instructional strategy and the sequence of concepts. Course objectives should always be stated so that an evaluation of the lesson has a concrete basis. All three types of objectives should be included in the lesson plan. A rationale helps the teacher decide why the material is important or what is to be gained from a discussion of this particular topic. The general objective allows the teacher to relate the specific activities of the day to a unit of instruction regardless of the length of the unit. The specific performance objectives detail what the student is to accomplish from the day's instruction. The preparation of the performance objectives has been discussed in detail and the teacher is referred to Part 4 for the mechanics of writing acceptable objectives.

The instructional strategy selected is determined by many variables. The typical strategies used are: lecture, laboratory, case history, and demonstration. Inquiry may or may not be present in any of these strategies. The amount of student involvement in topic development depends on the confidence of the teacher and his willingness to work with student ideas. Examples of lesson plans utilizing these strategies are included in this chapter.

Each class period requires a procedure for focusing the students' attention on the lesson material that needs to be formulated in advance. The questions that the teacher asks must be written out so that the anticipated answers may be examined. The quality of the question asked will determine the amount of interaction with the students and the level of understanding that they will seek. A formulated list of questions will keep the class discussion or other work oriented to achieving the stated objectives. Summary questions should be used to solidify ideas and generate new concepts to be investigated. The teacher should not use this portion of the class time to lecture on the concept or attempt to close related areas of investigation. If science is not a closed subject, then each individual concept is not closed; and the instruction should reflect this premise.

Assignments are varied and range from reading, to writing, to problem solving. Each topic will suggest a type of assignment, and the classroom interaction will often provide other resources to consult. Each assignment should be reasonable. The student should have a high probability of success and should be able to do the assignment in a reasonable length of time. If the time is too long or the assignment is too difficult, the student will not do the assignment and part of the learning opportunity will be lost.

All teaching supplies, including audiovisuals and laboratory equipment, should be listed. When lesson plans are made in advance, this list will enable the teacher to enter the classroom with all of the materials necessary to conduct that class. Films may be ordered, audiovisual equipment can be reserved, and other visuals may be produced. Chemicals, glassware, or related demonstration equipment can be set up and tested in advance.

Reference books for the student and the teacher should be listed. Magazine articles can also be noted for the students' use. The location of such materials can be noted so that they will be available when necessary.

Some time during the use of this lesson plan and preferably throughout the plan, the test questions should be developed that will correspond to the stated objectives. Some questions are formed when the objectives are written. Other questions will be raised from classroom interaction. Tests are easier to construct when the teacher prepares the questions as the instruction progresses. Some teachers write the questions and then teach the students so they will be able to answer them. Other teachers indicate that the best procedure is to examine the instruction after it has been given and then develop the questions based on student performance. Each procedure has its advantages.

Lesson plans are not permanent. A lesson plan should be used once and then revised. Space should be allowed on the lesson plan to write in teacher and/or student impressions of the success or failure of any individual part of the plan. Criticism must be given shortly after the use of the lesson plan or the little details will be forgotten by the teacher. Missing questions or good ideas may be noted and included on revisions for future use. The lesson plan may be a failure and not used again but should be included in the teacher's file for future reference. An explanation for the failure should be included for future revisions. A final evaluation of each daily lesson plan should be conducted at the end of each unit of instruction. The weak lesson plan should be marked for revision and the revisions should be made immediately. Good lesson plans should be labeled and reasons for their probable success noted. Success or failure of a daily lesson will often be the result of unexpected variables, which should be noted when they are recognized.

Lesson Plans

The following lesson plans are presented as models for various teaching approaches. It must be emphasized that they are examples, not exemplars. As examples, they have deficiencies and are open for modification and improvement. However, teacher-made lesson plans, including yours, will fit the same description. The lesson plans serve both as a guide to future work and as a record of what has been done. Last year's lesson plans can be used as a foundation for planning this year's lessons, and to do otherwise would be inefficient and would fail to utilize the experience acquired from their previous use. However, to use last year's plans for this year's lessons with no further consideration of them can be worse than inefficient. This can be ineffective.

Perhaps it would be best to regard these lesson plans as "last year's plans." They can be a basis for planning for your current lessons, although they might require modification. They are unlikely to be ready for immediate use in the classroom in their present condition.

Lesson Plan: Demonstration-Discussion

Topic: Conservation of Energy

General Objectives:

1. The student shall be able to incorporate the previously defined concepts of work, kinetic energy, and potential energy into the principle of conservation of energy.

2. The student shall be able to observe the behavior of isolated systems and to analyze this behavior in terms of the principle of conservation of energy.

3. The student shall be able to recognize the assumptions that must be satisfied before conservation of mechanical energy can be said to exist in a physical system.

Specific Objectives:

1. For a given physical system, the student shall be able to indicate why the expression:

$$\text{Energy}_{\text{initial}} = \text{Energy}_{\text{final}}$$

implies that *no* net work has been done on or by the system between the initial and final conditions. Acceptable performance would consist of analyzing the meaning of work and energy and stating why the condition of no net work done on or by the system must be satisfied if this expression is to be valid.

2. The student shall be able to apply the principle of conservation of energy in an analysis of interactions within isolated systems. Acceptable performance would include identification of the system elements that are involved in energy transfers within the system and an indication of which elements of the system gain or lose energy in these interactions.

3. Given data for an interaction in an isolated system, the student shall be able to perform a quantitative analysis of this interaction under the assumed conditions of conservation of energy.

4. The student shall be able to identify systems and situations in which the conservation of mechanical energy is not applicable. Acceptable performance would include stating whether the system's mechanical energy was increasing or decreasing and where this energy is coming from or going to in this interaction.

Hypothesized strategies for attaining these objectives:
1. Observation of a simple pendulum with a bob of rather large mass. (Resources required: 3 m of nylon rope, a bowling ball with attached screw-eye, a secure ceiling support, and a ladder)
2. Galileo's simple pendulum with a nail directly below the support point and halfway between the support point and the bob. (Resources required: 1 m of string, a pendulum bob or weight, and a pendulum support with rod or nail projecting out 0.5 m below support)
3. Bouncing super ball. (Resources required: a super ball and polaroid photographs or transparencies of its motion)
4. 1 kg mass bouncing on the end of a spring. (Resources required: 1 kg mass, a coil spring, a meter stick, and polaroid photographs or transparencies of this motion)
5. Hot wheels car on roller coaster track. (Resources required: hot wheels car, a track, photocells, and a timing circuit)
6. Project Physics film loops. (Resources required: the Aircraft Takeoff loop, the Pendulum and the Piledriver loop, a loop projector, and a screen)
7. PSSC film. (Resources required: Film *Energy and Work*, a 16mm sound projector, and a screen)

Selection of a strategy for attaining the objectives:
1. Resources available: 3 super balls, a set of weights, assorted coil springs, meter sticks, 1 bowling ball with attached screw-eye, 3 m of nylon rope, a secure ceiling support, a ladder, a film loop

projector, a screen, the Aircraft Takeoff loop, and the Pendulum and the Piledriver loop.
2. Criteria for selection:
   a. The required resources are available.
   b. The demonstrations should hold the students' interest and should be clearly visible.
   c. A transfer between several different kinds of energy should be observed.
   d. The strategy should lead to class discussion and quantitative analysis by the students.
3. Strategy selected:
   Number 1 will be used along with parts of numbers 3 and 6.

Procedure:
1. Opening presentation: Hold a super ball about one meter above the table and release it. Catch the ball on the first bounce without making the fact that the ball did not return to its initial height too obvious. Analyze the interaction in the earth-ball system, assuming that energy is conserved. Emphasize the gravitational potential energy—kinetic energy—elastic potential energy—kinetic energy—gravitational potential energy transfers that occur as the ball falls, bounces, and then rises again.

Diagram the motion of a simple pendulum upon the blackboard or have a transparency prepared (Figure 4-1).

The usefulness of the energy conservation principle should be emphasized by pointing out that the speed of the bob at the bottom, or at any other position, can be computed once the height of this position is known. For example,

$$\text{Total System Energy relative to the lowest point of the path} = \text{Initial PE} + \text{Initial KE}$$

$$= mgh_1 + 0$$

Thus at the bottom,

$$PE + KE = mgh_1$$

or

$$0 + \tfrac{1}{2}mv^2 = mgh_1$$

Hence at the bottom of the path,

$$v = \sqrt{2gh_1}$$

118 Additional Resources for the Teacher of Chemistry and Physics

An analysis of why the tension force in the pendulum cord does no work may be profitable since an explanation requires the use of the definition of work.

Use energy conservation arguments and show that the bob can rise no higher than h_2 at the right end of the path and will rise no higher than its initial height, h_1 upon its return swing to the left.

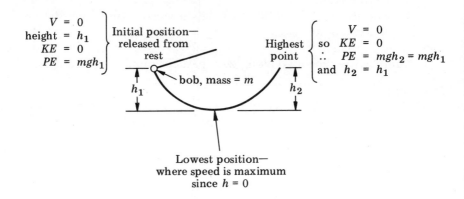

Figure 4-1. Sketch of a simple pendulum for energy analysis.

2. Demonstration: Have a bowling ball hung from 3 m of nylon rope attached to a secure support in the ceiling. The hanging of this pendulum bob could serve as a mind-capture device if demonstration number 2 precedes presentation number 1. Hold the pendulum bob up to the nose and then release the bob. Stand still and let it swing away and then come back. Student volunteers may also be used for this. The subject must stand still so that all can see that the final height is *less than* h_1.

3. Discussion questions:
   Is mechanical energy conserved?
   Where could the energy go?
   What would we have to do to get the pendulum bob back up to h_1?
   If conservation of energy doesn't work in a "real" situation, what good is it? (This question may arise earlier if the students note that the super ball does not return to its original height.)

   Use the film loop, *Conservation of Energy—Aircraft Takeoff*, either as another approach to this general problem or as a means of determining the students' mastery of the objectives. If further instruction seems necessary upon the basis of the discussion following the demonstrations, use the film loop *Conservation of Energy—The Pendulum and the Piledriver* as an evaluative measure.

Reference materials:
All students: PSSC Physics. Chapters 24 and 25. This will be the last day on Chapter 24 and will serve as an introduction for Chapter 25.
Alternate references:
Project Physics, Chapter 10.
Rogers, Eric, *Physics for the Inquiring Mind*, Chapter 29.
Teacher: *PSSC Teacher's Resource Book and Guide*, parts 24 and 25.
Hittle, Stekel, Stekel, and Andersen, *Sourcebook for Chemistry and Physics*, Chapter 1.

Materials and teaching aids:
Super ball, *secure* ceiling support for pendulum, 3 m of nylon rope, and bowling ball with eye attached for use as pendulum bob.
Film loop projector
Film loops: *Conservation of Energy—Aircraft Takeoff*
Conservation of Energy—The Pendulum and the Piledriver

Test item for evaluation:
1. A ball is thrown up into a large culvert that slopes up and into a bank. You threw the ball up into the culvert with a speed V but it comes back out at you with a speed greater than V. Which is the best explanation?
    a. This is the expected result that should occur as the ball goes in, bounces up the slope, slows down and stops, and then rolls back down and out toward you.
    b. This could occur only in a perfectly elastic collision between the ball and a partition inside the culvert.
    c. Someone is hiding inside the culvert and has caught the ball and thrown it back at you.
    d. The ball comes out going faster because it lost some energy while bouncing around inside the culvert.

Redirection:
How did the students respond to the initially overly simplified presentations that were shown to be inadequate?

Are more examples, from chemistry or meteorology or nuclear physics, needed to reemphasize the applicability of the principle of conservation of energy? Can this point be made adequately in the discussion of this principle and the demonstrations?

What fraction of the class was involved in some verbal interaction during the discussion?

Can the students apply conservation of energy, both qualitatively and quantitatively, in future discussions?

Lesson Plan: Laboratory
Topic: Uniform Motion

General objectives:
1. The student shall be able to make measurements and collect data in a laboratory situation.
2. The student shall be able to estimate the uncertainty in measurements and consider these uncertainties in the interpretation of experimental results.

Specific Objectives:
1. The student shall be able to collect data which will provide a record of an object's position at regular time intervals. Acceptable performance would include obtaining reproducible data which yield clearly separable locations for the object during its motion.
2. Given a record of the positions of an object at equal time intervals, the student shall be able to measure the necessary distances to obtain the data needed for a comparison of the velocity of an object during its motion. Acceptable performance would include identifying the motion as uniform or nonuniform according to the data and computations. All results should be within the uncertainties allowed by using ± one-half scale division as the minimum uncertainty for each measurement.
3. The student shall consider the uncertainties in the data when interpreting the results. Acceptable performance would include the use of these uncertainties in the discussion of the object's motion and its identification as uniform or nonuniform as described.

Hypothesized strategies for attaining these objectives:
1. Moving objects with data collected by multiple image photographic techniques. (Resources required: moving objects as determined by the description below, ASA 3000 Polaroid film, one or more Polaroid cameras, and a motor strobe or Xenon strobe or blinky.)

Moving Object	Motor Strobe	Xenon Strobe	Attached Blinky
battery-operated toy	x	x	x
dry ice pucks	x	x	
pucks on smooth surface and beads	x	x	
gliders on an air track	x	x	x
pucks on an air table	x	x	
dynamics carts with good wheels	x	x	x
human being (walking or running)	x		x
automobiles on street	x		

2. Synchronous spark timer and waxed paper tape providing a record of positions at 1/60 sec intervals. (Resources required: a rather fast-moving object with some type of an attached electrode, a synchronous or multifrequency spark timer, a paper tape, and a grounded conductor that the moving object passes at a uniform distance of about 1 mm)

3. Analysis of film loop data. (Resources required: a loop projector, *Analysis of Hurdle Race, I and II*, film loops.)

4. Use of a movie editor to analyze movies of objects in motion. (Resources required: a movie editor and a film of moving objects)

Selection of a strategy for attaining the objectives:

1. Resources available: Polaroid camera, a tripod, ASA 3000 Polaroid film, 2 battery-operated bulldozers with batteries, 12 dynamics carts, an air track with air source, 2 blinkies, 24 fifteen-centimeter rulers, and people.
2. Criteria for selection:
    a. The required resources are available.
    b. The data collection procedure is sufficiently reliable.
    c. The student should be able to see both the moving object and a permanent record of its position at uniform time intervals.
    d. The data collection procedures learned from this laboratory experience should help the student with later laboratory work.
    e. This activity should lead into a group discussion at the end of the period.
3. Strategy selected:

Number 1, using Polaroid photography of blinkies attached to the following moving objects: toy bulldozer, dynamics cart, air track glider, and a person.

Procedure:

1. Prelab discussion: Run the toy bulldozer across the table top; the students should not have seen it before. Ask, "What do we need to measure to determine if motion like this is uniform?"

Emphasize operational definitions for motion and uniform, relating the first to average velocity and the second to the intrinsic uncertainties in measurements.

Describe operation of Polaroid camera and blinky.

Divide the class into eight groups as follows: two groups each to the bulldozer, dynamics carts, gliders on the air track, and just walking or running. Each group should get their own picture of their activity. Dynamics cart and air track groups should consider how to get their objects' motion as uniform as possible while photography is begun by the other groups.

Groups must know what they want to photograph before using the camera. Keep the camera in use and avoid the monopolizing of it by any individual or group.

2. Postlab discussion: There were four different objects whose motion was studied. Is everyone getting the same uncertainty in measurement? Relate this to the rulers that all of the groups are using.

Although the time between flashes is not known, what did we have to assume about the flashes in order to *compare* the motion of the objects as they moved? Is there any way that we could test this assumption?

Have the groups report on their results. Are the groups working with the same object consistent with each other? Note whether any of the motion was nonuniform or which was least likely to be uniform on the basis of the data. Discuss this and any other differences.

References:
Student: Project Physics *Student Guide*, Experiment E4
Teacher: Project Physics *Teacher Guide*

Resources required:
2 bulldozers with batteries, 2 dynamics carts, air track with air source and gliders, 2 blinkies, at least 1 (but preferably 2) Polaroid cameras and tripods, 2 or 3 packs of film, rulers with 1 mm scale divisions (at least one for each group).

Evaluation:
Attempt to interact with each group during their data analysis.

Have as many students as possible involved in the postlab reports and discussion.

Have the students hand in a written report. If inadequate understanding is evidenced here or in the discussions, have the students study and analyze the "Hurdle Race" film loops.

Have the students suggest alternate topics and/or procedures. Discuss some of these such as the use of motor or xenon strobes, or the movie editor analysis of movie films.

Redirection:
Does the small number of cameras create too much of a bottleneck in data collection? If so, would providing some prints for immediate analysis to half of the class while the other half starts with photography be a more efficient procedure?

Do any of the student comments or suggestions seem feasible for incorporation into this experiment?

On the basis of the students' work with uncertainties, what further instruction will be needed?

Lesson Plan: Case Study

Topic: Electromagnetic Induction

General Objectives:
 1. The student shall be able to describe the role of experimental results and the role of earlier hypotheses in each stage in the formulation of physical theories.
 2. The student shall be able to recognize the importance of the careful planning and extensive work that is required in the development of theories, even though this type of work is often overshadowed by breakthroughs. Acceptable performance would include an analysis of the conditions that led to the breakthrough and a description of how this new information was incorporated into a theoretical structure.
 3. The student shall be able to recognize the multiple approaches that are followed in the development of a theory. Acceptable performance would include stating examples of simultaneous investigations that were performed independently by different investigators and evaluating the results of these investigations to arrive at a rational determination of which of these investigators should have received credit for the discovery. Shared credit should be allowed and even encouraged.

Specific Objectives:
 1. The student shall be able to indicate the general conditions under which a current will be induced in a conductor. Acceptable performance will include describing the role of the magnetic field and the changes in the field in the student's own words.
 2. The student shall be able to relate the investigations of Oersted, Faraday, Henry, and Lenz to the development of the theory of electromagnetic induction. Acceptable performance will include identifying at least one contribution to the theory of electromagnetic induction from the work of each of these investigators.
 3. Given a magnet of known polarity and a coil of wire whose windings are clearly visible, the student shall be able to predict the direction of the induced currents that are caused by the insertion and removal of the magnet from the coil.

Hypothesized strategies for attaining these objectives:
 1. Library work: Individual students will be assigned to study various aspects of electromagnetic theory, its historical development, or its application and will deliver reports on their findings. (Resources required: history of science sources that can be read by the students, such as books like Holton and Roller; several days for the students to do this library work.)

2. Have students read original works of the investigators. (Resources required: original reports or abstracted or edited versions of these works that can be read by the students)

3. Demonstrations of electromagnetic induction in the sequence of their discovery and discussions of their role in the development of the theory of electromagnetic induction. (Resources required: sensitive compass, battery or power source capable of delivering a large current, multirange galvanometer, several meters of fine (18-22 gauge) wire, iron rod, copper and zinc plates, acid, strong bar magnet)

4. Film loops on electromagnetic induction. (Resources required: loop projector, loops from Commission on College Physics film repository, other applicable films)

Selection of a strategy for attaining the objectives:

1. Resources available: Project Physics test and readers, one copy of Holton and Roller, issues from the last few years of *Scientific American*, one set of Science Study Series paperbacks, 12 bar magnets, 2 rolls of No. 22 wire, 4 galvanometers, 8 compasses (6 are functional), many 1/2" bolts from the shop, 7 copper plates, 4 zinc plates, sulfuric acid, 12 volt automobile battery, film loop projector, 16 mm sound projector.

2. Criteria for selection:
    a. The required resources are available.
    b. The stages in the development of a theory, in this case electromagnetic induction, should be visible and amenable to analysis and discussion.

3. Strategy selected:

Number 3, using demonstrations in historical sequence, with discussion supplemented by outside reading.

Procedure:

Repeat Oersted's demonstration showing the effect of a current flowing through a wire upon a compass held near the wire. Emphasize Faraday's contribution: the recognition of the circular pattern around the wire, and the introduction of the field concept itself.

Discuss the next step as suggested by symmetry: Since a magnetic field is produced by an electric current, shouldn't an electric current be produced by a magnetic field?

Demonstrate the null reading that results when wires are simply wrapped around a magnet and are then connected to a current detector. Use current detection procedures of increasing sensitivity: first feel the wire to see if it gets hot, and then use a galvanometer.

Replicate Henry's experiment (Figure 4-2). Insertion of plates into acid produces a deflection that subsides once the current is steady; opposite deflection is produced upon the removal of the plates—again subsides.

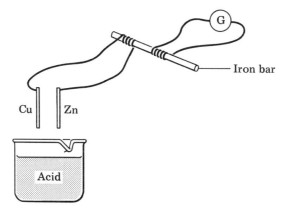

Figure 4-2. Apparatus arrangement for replication of Henry's experiment.

Describe the analogous experiments conducted by Faraday showing that induced currents are produced only by changing currents. Emphasize that these experiments were done independently.

Emphasize that changing currents induce changing magnetic fields. Discuss how changes in the magnetic field strength can be produced by using the permanent magnets as a source of the magnetic field.

Demonstrate by moving a magnet into and out of a coil of wire that is connected to a galvanometer. Note the pattern that is evident in the induced currents and the changes in the field strength.

See if the students will arrive at something similar to Lenz's law on the basis of these observations. Point out how this relationship encompasses all of the phenomena that were previously encountered in electromagnetic induction.

References:
Student: Project Physics, Chapter 15
Additional or optional references for students:
  MacDonald, *Faraday, Maxwell and Kelvin*
  Holton and Roller, *Foundations of Modern Physical Science*, Chapter 28.
Teacher:
  Project Physics *Teacher Guide*
  PSSC Physics, Chapter 31
  Holton and Roller, Chapter 28

Redirection:
Would the presentation of some of these demonstrations by a student investigator or a team responsible for setting up and showing the particular phenomena be more effective than teacher demonstrations?

How effective were the references in Holton and Roller and in MacDonald's paperback?

Should some of the work with the magnet being inserted and removed from a coil of wire be done as a laboratory exercise instead of a demonstration?

Lesson Plan: Laboratory
Audience: Eleventh Grade Students
Topic: Development of the Periodic Properties and the Periodic Chart

General Objectives:
1. The student shall develop the ability to take data, classify it, and draw conclusions from it.
2. The student shall develop an understanding of the periodic properties of the elements.
3. The student shall develop an understanding of how the elements were originally classified.

Specific Performance Objectives:
1. If the student is given a list of general formulas, such as X_2Y_3, he shall be able to tell to which family of elements the X and Y belong on the basis of oxidation number alone. Acceptable performance would consist of being correct for eight out of ten elements.
2. If the student is given a list of formulas for binary compounds, he shall be able to classify the formulas on the basis of the combining capacity of one of the elements in the compound. Acceptable performance would include being 100 per cent correct on the basis of his classification.
3. If, in a laboratory situation, the student is given the atomic weights and several representative formulas including each element, he shall be able to devise some type of classification and be able to justify it orally when quizzed by the instructor. Acceptable performance would include a meaningful order, such as oxidation state of the anion, with a minimum of two stated periodic trends in this system.
4. In this laboratory, when given the list of compounds to classify, and after they have worked with the list, the students shall be able to predict three of the following.
    a. There should be some elements with "0" combining capacity.

b. There should be a column of elements between the column headed by lithium and the column headed by fluorine.
   c. There should be eight to ten elements between calcium and gallium.
   d. There should be twenty to twenty-four elements between barium and thallium.

Approach:

At the beginning of class have a number of differently shaped and colored geometrical solids with different numbers of nails sticking out of them. Start discussing the different ways that the solids could be classified.

Development:

1. Hand out the list of compounds and atomic radii and explain that the students are to classify the elements from these compounds in as many different ways as possible and to defend their classifications. Ask them what other data they would like to have for these elements. After several elements have been listed, allow the students to use the atomic weights from their books.*

2. After twenty minutes, have the students assemble in groups of four to share their classification systems and develop a classification(s) system as a group.

3. After another ten minutes, have a volunteer from each group list the elements on the board as their group has classified them.

4. Ask the class, in general, which method (or methods) seems to be most logical. Allow discussion from the class and, as much as possible, have the proponents of each method defend their classification schemes. The class should now be ready to define a good classification system.

5. Is there anything these schemes have in common? It is hoped by this time that the students will be able to recognize many of the periodic properties of elements. If not, give hints until specific performance objective number 4 is achieved.

6. It is important to develop questions based on the students' weak areas. It will thus be necessary to ask questions that seem pertinent at the time and cannot be foreseen in a lesson plan.

---

*If the students are unable to get started, give them the hint that they should have a homogeneous group of cations or anions.

128 Additional Resources for the Teacher of Chemistry and Physics

Instructions:
You are to take the following information and classify the elements in as many different ways as possible.

			Element	Atomic Radius
LiCl	CCl_4	$AsBr_5$		
NaF	SnO_2	BaF_2	F	0.64 Å
			O	0.66
$AlCl_3$	GeS_2	SrO	Na	1.86
			Te	1.37
$PbCl_4$	N_2O_5	$SnCl_4$	Ba	2.17
			Tl	1.71
Rb_2O	AsI_5	PI_5	Hg	1.21
			Cu	1.28
BeI_2	BaTe	Tl_2Te_3	Bi	1.46
			Al	1.43
Li_2Te	CdF_2	SrF_2	Br	1.14
			Rb	2.44
CH_4	P_2O_5	MgO	Sb	1.41
			S	0.88
$SiCl_4$	$BiCl_5$	$GaBr_3$	N	0.70
			Sr	2.15
PbO_2	Rb_2S	In_2Se_3	P	1.10
			I	1.33
MgS	$GeBr_4$	SiO_2	As	1.21
			Ca	1.97
BeSe	CuI_2	CO_2	K	2.31
			Li	1.52
$HgCl_2$	Na_2Se	$TlCl_3$	S	1.04
			Cs	2.62
AgBr	Tl_2O_3	Al_2Te_3	Pb	1.75
			Sn	1.40
KF	$SnBr_4$	Cs_2O	Ge	1.22
			Mg	1.60
$InCl_3$	N_2Te_5	MgI_2	Si	1.17
			Se	1.17
GaBr	Ga_2Se_3	$SbCl_5$	C	0.77
			Ga	1.22
BI_3	$CaBr_2$	Sb_2Se_5	Be	.73
			In	1.62
K_2O	CsI			

Summary Questions:
1. What would the criteria be for designing a good classification system?
2. What information did we have about these elements?
3. Would the other properties of elements that we mentioned at the start of class make any difference in our system?

Student and Teacher References:
Metcalfe, H. C., J. E. Williams, and J. F. Castka, *Modern Chemistry* (New York: Holt, Rinehart, and Winston Inc., 1966), pp. 64-78.
Choppin, G. R., and B. Jaffee, *Chemistry* (Morristown, N.J.: Silver Burdett Company, 1965), pp. 170-184.
Boyland, Paul J., *Elements of Chemistry* (Boston: Allyn & Bacon Inc., 1965), pp. 130-149.
Smoot, R. C., J. Price, and R. L. Barrett, *Chemistry: A Modern Course* (Columbus: Charles E. Merrill Publishers, 1965), pp. 120-141.

Assignments:
For tomorrow I would like you to have some answers to the following questions:
1. From your text and/or other chemistry books, compile a list of all the properties that have been used to classify elements.
2. Given the compound X_2O_5, what can you tell about X? What would the formula for the compound of X and chlorine be?
3. How do we fit compounds such as C_2H_6, C_2H_4, and C_2H_2 into our classification scheme?

Lesson Plan: Demonstration-Discussion
Audience: Eleventh Grade Students
Topic: Acids and Bases

General Objective:
To acquaint the student with the concepts of acids and bases.

Specific Performance Objectives:
1. The student shall be able to describe in writing three acid-base concepts as discussed in class. This discussion should include the names associated with each concept, a basic statement of each concept, an example of a compound or substance that is an acid and one that is a base according to each concept, and an equation illustrating the acid-base reaction of each concept.
2. Given a list of compounds, the student shall be able to classify each compound as being either an acid, a base, both acid and base, or neither according to the Arrhenius, Bronsted, and Lewis concepts. The student shall be able to include the equations to illustrate the

reason for his classification and for the Lewis concept. He should be able to include the electron-dot configuration of the compound.

3. Given the equation for the reaction of a Bronsted acid and base, the student shall be able to select and indicate which substances are conjugate acids and which are conjugate bases. Given the predominant direction of the reaction, he shall be able to conclude and indicate in writing the relative strengths of the conjugate acids and the conjugate bases of the reaction.

4. The student shall be able to illustrate by writing equations how water can function as either an acid or as a base according to the Bronsted concept.

Procedure:
1. List on the board a number of substances, for example, HCl, HNO_3, HCN, KOH, $Ca(OH)_2$, HF, NaOH, and NaCl.
2. Establish the idea of these substances being electrolytes by asking the students what would happen if you were to put one of the substances, HCl, for example, into solution and were to test the solution with the conductance equipment. (Students are familiar with this experiment.)
3. Review with the students the fact that litmus paper indicates whether a solution is acidic or basic.

Development:
1. Test solutions of HCl, NaCl, and NaOH, having the students observe and discuss the results. Ask them why the results are acidic, neutral, and basic to litmus test. Have the students identify ions in solution in each instance.
2. Test additional solutions (HNO_3, KOH, for example) and discuss all the substances listed on the board if necessary to have the students conclude that the acids are those substances that ionize in solution to give H^+, whereas bases are those that ionize in solution to give OH^- in solution.
3. Point out to the students that this is the Arrhenius concept of acids and bases.
4. Ask the students if $SO_2(OH)_2$ is a base, then rewrite the compound as H_2SO_4 to illustrate that one cannot always tell by inspection of the formula if a substance is acidic or basic.
5. Discuss the Arrhenius concept extended to "increasing the H^+ or OH^- concentration" in solution:
    a. Water ionizes in solution to give hydrogen ions and hydroxyl ions,

$$H_2O \rightleftharpoons H^+ + OH^-$$

b. However, the situation is known to be that each of these ions is hydrated, therefore the more accurate description is,

$$\text{H:}\overset{xx}{\underset{\underset{H}{..}}{\text{O:}}} + \text{H:}\overset{xx}{\underset{\underset{H}{..}}{\text{O:}}} \rightleftharpoons \text{H:}\overset{xx}{\underset{\underset{H}{..}}{\text{O:}}}\text{H}^+ + \text{:}\overset{xx}{\underset{xx}{\text{O:}}}\text{H}^-$$

c. For simplicity, however, the dissociation or ionization is written as,

$$\text{H}_2\text{O} \rightleftharpoons \text{H}^+ + \text{OH}^- \qquad \text{(slight)}$$

d. Thus, for an acid in solution, for instance, the more accurate description is,

$$\text{HCl} + \text{H}_2\text{O} \rightarrow \text{H}_3\text{O}^+ + \text{Cl}^-$$

in which the H_3O^+ is the hydronium ion.
6. Develop the full implications of this extended concept.
   a. With a small beaker of water and a piece of glass tubing, blow into the water and test the resulting solution with litmus.
   b. Have students conclude that:

$$\text{CO}_2 + \text{H}_2\text{O} \rightarrow \text{H}^+ + \text{HCO}_3^-$$

thus showing that the hydrogen increase may be brought about by a substance that does not itself contain hydrogen.
   c. Similarly, show that there can be cases that do not contain OH⁻ by a discussion of the reaction,

$$\text{NH}_3 + \text{H}_2\text{O} \rightarrow \text{NH}_4^+ + \text{OH}^-$$

   d. Furthermore, have the students realize that this definition limits one to aqueous solutions by having them notice that it is the H⁺ and OH⁻ balance of H_2O that is being effected.
7. Discuss concepts that were developed to apply to nonaqueous, as well as aqueous, systems.
   a. Discuss the Bronsted concept.
      1) By definition, *acid* refers to any proton donor and *base* refers to any proton acceptor, even in nonaqueous solvents.
      2) The generalized equation to relate acid and base is

$$\underset{\text{acid}}{\text{HA}} \rightleftharpoons \underset{\text{proton}}{\text{H}^+} + \underset{\text{base}}{\text{A}^-}$$

3) For an acid to act as a proton donor, a proton acceptor must be present to receive the proton. In aqueous solution, the proton acceptor is H_2O.

$$HA + H_2O \rightleftharpoons H_3O^+ + A^-$$
$$\text{acid}_1 \quad \text{base}_1 \quad \text{acid}_2 \quad \text{base}_2$$

4) Since H_3O^+ can also act as a proton donor if the reaction goes to the left, it is termed the conjugate acid of the base H_2O; similarly A^- is the conjugate base of the acid HA.
5) In a nonaqueous solution an example would be,

$$HCL + NH_3 \rightarrow NH_4^+ + Cl^-$$

6) Base and acid strength is determined by a comparison of conjugate bases with bases and conjugate acids with acids knowing the direction of the reaction.

b. Discuss the Lewis concept.
 1) An acid is any species (molecule or ion) that can accept a pair of electrons; a base is any species (molecule or ion) that can donate a pair of electrons.
 2) Considering the reaction

$$H_2O + HCl \rightarrow H_3O^+ + Cl^-$$

notice that an acid-base reaction in the previous concepts is also an acid-base reaction in the Lewis concept:

$$H\!:\!\ddot{\underset{H}{O}}\!: + H\!:\!\ddot{Cl}\!: \rightarrow [H\!:\!\ddot{\underset{H}{O}}\!:\!H]^+ + [:\!\ddot{Cl}\!:]^-$$

According to the Lewis concept, the oxygen atom donates the electron pair and the H^+ ion of the HCl accepts the electron pair.
 3) Furthermore, the more general application of the Lewis concept can be seen in reactions in which the Arrhenius and Lowry—Bronsted concepts are not applicable.

$$2\ H\!:\!\underset{H}{\overset{H}{\ddot{N}}}\!: + Ag^+ \rightarrow \left[H\!:\!\underset{H}{\overset{H}{\ddot{N}}}\!:\!Ag\!:\!\underset{H}{\overset{H}{\ddot{N}}}\!:\!H\right]^+$$

Summary:
How would you (the student) classify the following reactions or particles according to acids or bases in each of the concepts?

1. $CaO + SO_3 \rightarrow CaSO_4$

   $BF_3 + F^- \rightarrow BF_4^-$

   $NH_3 + H_2O \rightarrow NH_4^+ + OH^-$

   $NH_2^- + H_2O \rightarrow NH_3 + OH^-$

   $HCl + NH_3 \rightarrow NH_4^+ + Cl^-$

2. HSO_4^-, H_2S, CO_3^{-2}, NH_3, NH_4^+, OH^-, CH_3NH_2, HCO_3^-, H_2O, $Cu(H_2O)_4^{+2}$, $Fe(H_2O)_5OH^{+2}$.

Resources:
- Litmus paper
- Distilled water
- 1 hollow glass tubing
- 0.5 g. NaCl
- Eyedroppers
- 50 ml. or 100 ml. beakers
- 5-10 ml. of stock solutions of selected acids and bases such as HCl, HNO_3, NH_4OH, KOH, $NaOH$, H_2SO_4, $HC_2H_3O_2$.

The number of eyedroppers, beakers, and solutions may be varied; minimum necessary is 4 beakers, 3 eyedroppers, and 3 solutions.

Assignment:
Read in Smoot pages 314-320 and write out problems 1 and 2 on page 333.

Student References:
Choppin, Gregory R., and Bernard Jaffe, *Chemistry: Science of Matter, Energy, and Change* (Morristown, N.J.: Silver Burdett Company, 1965), Chapter 17.

Posin, Dan Q., and Marc Shampo, *Chemistry for the Space Age* (Philadelphia: J. B. Lippincott Co., 1964), Unit Six.

Sienko, Michell J., and Robert A. Plane, *Chemistry*, 2d Ed. (New York: McGraw-Hill Book Company, Inc., 1961), Chapter 10, Section 8.

Smoot, Robert C., et al., *Chemistry, A Modern Course* (Columbus, Ohio: Charles E. Merrill Publishers, 1970), Chapter 16.

Teacher References:
Choppin, Gregory R., and Bernard Jaffe, *Chemistry: Science of Matter, Energy, and Change* (Morristown, N.J.: Silver Burdett Company, 1965), Chapter 17.

Nebergall, William H., et al., *General Chemistry*, 2d Ed. (New York: McGraw-Hill Book Company, Inc., 1961), Chapter 10, Section 8.

Smoot, Robert C., et al., *Chemistry, A Modern Course*, Teacher's Annotated Edition and Solutions Manual (Columbus, Ohio: Charles E. Merrill Publishers, 1970), Chapter 16.

Lesson Plan: Lecture-Demonstration
Audience: Eleventh Grade Students
Topic: Physically Observable Properties of a Chemical Reaction

General Objectives:
 1. The demonstrations shall acquaint the student with the more common physically observable properties of chemical reactions.
 2. The student shall be given the opportunity to observe and interpret his observations.
 3. By using seemingly contradictory examples the student shall improve his powers of observations.

Specific Performance Objectives:
 1. Given any one of the demonstrations, the student shall be able to state what he has seen and why he has seen it. Acceptable performance will include two oral observations and two logical reasons for them.
 2. The class shall be able to make observations and to interpret the data orally. Acceptable performance will include the listing, on the board, of all the physically observable properties from the demonstration, and the discovery that all of these chemical reactions involved the release of energy.
 3. After observing the demonstrations and participating in the discussions, the student shall be able to observe the auxiliary demonstrations and tell which are chemical reactions. Acceptable performance will include identifying 80 per cent of the demonstrations correctly and being able to write two justifications for each of his choices.

Approach:
 Start the class with an egg frying in a skillet. Ignore all questions for several minutes and then start the discussion.
 1. What is happening in the skillet?
 2. How did you observe that an egg was frying?
 3. What happened to the egg?
 4. Is there a chemical reaction taking place in the skillet?
 5. How do you suppose we can detect a chemical reaction?

There will undoubtedly be a number of questions that the students will raise. These should be answered or directed to another student for answering unless the question is irrelevant.

Development:
I. A gas is evolved.
   A. Place a piece of zinc in a test tube of sulfuric acid.
      1. What did you observe?
      2. What caused it?
      3. Is this a chemical reaction?
   B. Shake a warm Coke and remove the cap.
      1. What did you observe?
      2. What caused it?
      3. Is this a chemical reaction?
II. A precipitate is formed.
   A. Place a solution of silver nitrate and sodium chloride in a beaker.
      1. What did you observe?
      2. What caused it?
      3. Is this a chemical reaction?
   B. Cool a saturated potassium nitrate solution.
      1. What did you observe?
      2. What caused it?
      3. Is this a chemical reaction?
III. Light is emitted.
   A. Ignite a strip of magnesium ribbon.
      1. What did you observe?
      2. What caused it?
      3. Is this a chemical reaction?
   B. Turn on a light bulb.
      1. What did you observe?
      2. What caused it?
      3. Is this a chemical reaction?
IV. Sound is emitted.
   A. A firecracker is exploded.
      1. What did you observe?
      2. What caused it?
      3. Is this a chemical reaction?
   B. A balloon is punctured.
      1. What did you observe?
      2. What caused it?
      3. Is this a chemical reaction?

V. Color change.
   A. Pour sulfuric acid into a beaker containing sugar.
      1. What did you observe?
      2. What caused it?
      3. Is this a chemical reaction?
   B. Place one drop of $KMnO_4$ in a large volume of water.
      1. What did you observe?
      2. What caused it?
      3. Is this a chemical reaction?

Again many of the discussions will depend on the types of questions that the student will have.

Summary Questions:
1. How many ways can we detect a chemical change?
2. How much evidence do we need to decide?
3. Why is it easy to be fooled into thinking that a reaction has taken place?

Materials:

Equipment	*Chemicals*
Bunsen burner	zinc
tongs	water
beakers	sulfuric acid
test tubes	sodium chloride
balloons	silver nitrate
light bulb and socket	potassium nitrate
test tube rack	potassium permanganate
test tube holder	sugar
asbestos pad	firecracker
ringstand	sand
iron ring	
wire gauze	
goldfish bowl	

Demonstrations to be Used for Testing:
1. Cooling a solution of copper sulfate
2. A candle burning
3. Calcium chloride and ammonium carbonate in solution
4. Allow a mixture of water and mud to stand
5. Place dry ice in water
6. Observe a baking cake

7. Pop a paper cup
8. Burn ammonium dichromate
9. Place some I-KI solution on a piece of bread
10. Observe a phosphorescent compound in the dark

Teacher References:

Smoot, R. C., J. Price, and R. L. Barrett, *Chemistry: A Modern Course* (Columbus, Ohio: Charles E. Merrill Publishers, 1965), pp. 60-66.

Student References:

Boylan, Paul J., *Elements of Chemistry* (Boston: N. Allyn & Bacon Inc., 1965), pp. 241-256.

Eblin, Lawrence P., *The Elements of Chemistry* (New York: Harcourt, Brace and Jovanovich, Inc., 1965), pp. 7-8.

Assignments:

For tomorrow I would like you to formulate answers to the following questions:

1. What are other physically observable properties of chemical reactions? Why do they occur? How would you demonstrate them?
2. Is there any way to detect chemical reactions other than by measuring physical changes?

Textbooks and Programs

The textbook selected by the teacher for use by the students will be a major factor in the success or failure of the students. It is necessary for the teacher to approach this task of textbook selection with the purpose of finding the one that is most compatible with his style of teaching. Initially, there may be no choice for the teacher as school systems will often adopt books for periods of up to five years. Other limitations include those imposed by state boards of education that narrow the teacher's selection. New curriculum programs have been instituted in recent years and their efforts to produce a viable curriculum should be examined for the materials and ideas that they will bring into the classroom. The teacher must choose the best possible text or program in view of the impact that it will have on students.

Criteria need to be established by which programs and textbooks can be evaluated. Scales have been devised (Owen, Archie M., "Selecting Science Textbooks," *The Science Teacher*, 29:20-23, November, 1962) that allow the teacher to place a quantitative value on a particular item; and then, by summing all of these values, a rank order of the books evaluated may be determined. This is an idealistic approach to the problem, but it would remove part of the doubt from the teacher's mind as to whether he is right or wrong in the selection of a particular book. Although this procedure will indicate the book having the most overall positive points, it does not necessarily indicate the book that will produce the greatest benefit for the students.

Some of the factors that should be considered during a textbook evaluation are:

1. How will it be used?
2. How relevant is its content?
3. What is the quality of the questions and problems?
4. Is it readable?
5. Are the mechanics of the book adequate?
6. What supplemental materials are available?
7. What are the qualifications of the authors?

The text may be used as a supplement to the classroom activities or it may serve as the primary source of information. Some teachers use the text as an outline to the material by sequencing the class activities to coincide with the table of contents of the text. If the text is to be supplementary, then the material presented in it must be complete and understandable as well as interesting to the student. The students should be able to follow the presentations of the text with a minimum of teacher assistance. It should be well illustrated and the problems should be presented with several examples worked out and explained in detail. To evaluate these points, a selected group of former students should be asked to read representative portions of the text and present their comments. A teacher is not an adequate judge. If the book is to serve as the structure for the course, it must go into greater detail and leave little for the teacher to develop on his own. Ideally the textbook should include lesson plans. Consideration should be given to the fact that the longer the book is in use the more outdated it will become. Adopting a radically different text or program requires that it be viable for the length of its adoption. What seems good on paper does not always work well in the classroom.

The relevancy of the content will depend upon the training and prejudice of the teacher. Books are written that appeal to the factually oriented teacher. Texts are also geared to those who view the study of chemistry and physics as an open-ended activity. The content should be an accurate representation of the field when the text was written. Quantum mechanics, thermodynamics, and molecular orbital theory are often ignored because they are considered "too difficult" for the students to understand. A teacher who uses the book as a supplement will be able to make modifications and additions that will be suited to the students' abilities. It is easier for a teacher to delete items from the curriculum than to insert them.

The questions and problems given in a book are a guide to the information that is presented therein, and provide the students the direction needed to assimilate the information as they encounter it. The questions and problems provide a means of acquiring the skills of problem solving that are an integral part of chemistry and physics. For these reasons the questions should not be closed to the extent there is one simple answer. The student should begin with the simple questions, but more complex questions should also be available that require the student to analyze, synthesize, and evaluate. Problems should be presented in a direct and clear manner and should not be designed to trick the student. Answers should be contained somewhere in the book for the student to enable him to check his work to determine if he is acquiring the required problem-solving skills. A student who has recently had the course should be asked to solve specific problems from the text being evaluated, and if he experiences difficulty, the teacher should question the success that other students will have using that text.

A textbook should be readable. Although chemistry and physics students usually come from the upper half of the class, a text should not be selected that will test their reading abilities. Most ideas that are worth remembering can be expressed with two-syllable words. Sentence structure should also be simple as the complexity of the ideas presented may be compounded by the choice of phrases.

The mechanics of the book should be considered but not used as the final criteria for selection if all other things are equal. Among the factors to be considered are: organization, size of book, strength of binding, size of print, cover design, and quality of illustrations or photographs.

If the teacher searches extensively, he will find a book that sequences the course material satisfactorily. As a result of the equivalency of some concepts, authors may arrive at different orders for texts. The teacher should look beyond this point and determine if it is possible to rearrange the sequence of chapters of the book without confusing the students. This should be one of the prime criteria for the selection of a text based on mechanics.

Books come in all sizes, but this will usually make no more difference than the size or style of print used. As long as it is of sturdy construction, the text should be adequate. The size and style of the print should also make no difference as long as it is readable by all of the students.

The quality and quantity of illustrations and photographs are important. First, they provide the student with an example of what is being discussed and also provide a motivating force for continued reading. Second, illustrations or photographs provide a means of condensing a lengthy description of the relationship between variables, and the text and illustrations enhance one another with respect to the student's level of comprehension. The use of color in photographs and two or more colors in graphs to separate areas are other features that add to the quality of the text.

The availability of supplemental references and resources will strengthen any book or program. Greater inclusion of supplementary references usually reflects more effort by the authors to make the material understandable to the student. All or part of the following should be available: teacher's guide, answers to questions, appendices containing tables of information and "how to do it" information and "how to do it" information, standardized tests, a laboratory book that is compatible and complementary to the material of the text, film loops, slides, and the like. The relative quality of each should also be evaluated as availability and usability are not necessarily synonymous.

The qualifications of the author should be the last consideration. It should be assumed that if all other considerations are satisfactory, the author's academic achievement is unimportant. If any of the author's qualifications is to be considered, the teacher should determine if the author has ever taught a high school chemistry or physics course. An author who has taught at the high school level should be able to prepare materials usable at that level, since he has presumably experienced the same type of problems the teacher will.

The teacher is urged to take these areas of concern and formulate a list of specific points to consider. A scale of 1, for excellent, to 5, for unacceptable, should be entered after each point for all books evaluated. The teacher is also urged to assign a weight or importance factor to each question that will be used to multiply the scale value assigned to the question. This will allow the teacher to assign a greater importance factor to some questions than to others. The total scores should then be totaled and the books having the lowest two, three, or five scores can then be given further consideration. This is more useful in chemistry than in physics because of the greater number of chemistry texts.

Curriculum Projects

The curriculum projects that were undertaken by diverse groups of individuals in the 1960's and funded by the National Science Foundation will be examined in some detail. The teacher is urged to consider the programs, since much thought and effort was exerted by a variety of experts to produce these texts and the supplementary material. One other feature that curriculum projects share is that of developmental testing. Large numbers of classroom teachers not initially associated with the curriculum projects used the materials in their classrooms and provided feedback for improvements in the materials that were produced. The improvements were made and then retested to eliminate most of the problems that normally occur. Teacher handbooks were also prepared that provided assistance to the teacher in developing ideas and experimentation with the students. The specific projects to be examined include the Chemical Bond Approach Project, the Chemical Education Material Study, the Physical Science Study Committee, Project Physics, and the Engineering Concepts Curriculum Project.

Chemical Bond Approach

The Chemical Bond Approach (CBA) to teaching high school chemistry emanated out of a conference sponsored by the Division of Chemical Education of the American Chemical Society in the summer of 1957. The primary participants at the conference were a number of high school and college chemistry teachers who concluded that there must be improvement in the chemistry curriculum in the secondary school. The fulfilled objective of the program, they proposed, would prepare the student for an improved first year course in general college chemistry. The developers of CBA also realized that those students who were not planning to go to college should develop an understanding of chemistry and the chemical processes. One of the major premises was that high school students should study chemistry as an interrelation of facts and ideas. This would enable the students to gain an important perspective of the strategies and tactics used in chemistry.

The group of teachers that met to develop the materials for such a course arrived at the following objectives to guide them:

1. To construct a course that would serve as a prerequisite to the first-year college chemistry course.
2. To increase the student's critical thinking ability.
3. To present the topics of chemistry in an obviously logical order.
4. To develop a course that would be appealing to the average chemistry student as well as challenging to the more advanced student.

The initial writing took place in 1959. Seventeen individuals constructed the framework for the course and assisted in writing the materials to be used. The final selection of the material to be included in the text and the laboratory manual indicated the restriction of content to five areas:

1. The Nature of Chemical Change
2. The Electrical Nature of Chemical Systems
3. Models as Aids to the Interpretation of Systems
4. Bonds in Chemical Systems
5. Order, Disorder, and Change

To successfully achieve the project's objectives, the student would be placed in the laboratory investigating facts and concepts as opposed to listening to a teacher pouring forth a vast amount of factual information. The CBA materials emphasize the manner by which chemists arrive at knowledge through scientific inquiry. It also stresses both inquiry and the results of inquiry, methodology as well as recorded discovery.

The CBA material consisted of a textbook, an extensive laboratory manual, and two teacher's guides. The text and the laboratory manual are correlated to complement one another. To do this meant omitting much of the material that had been formerly included in typical chemistry texts. The textbook itself is a systematic course and should not be used as an elementary reference book. The laboratory theme is constructed upon the idea that the student develops a better understanding of chemistry by experimenting. The laboratory manual initiates a series of ten experiments, each posing a problem and nearly complete directions for solving the problem. As the student passes this point, he enters the second set of experiments that continue stating the problem, but these experiments are not as specific either in the directions to follow or the type of results to expect. In the third set of experiments found in the laboratory manual, the student is merely presented a problem to solve and must develop a laboratory procedure for collecting and then interpreting the data.

A typical laboratory session in the CBA program consists of a prelaboratory segment in which the teacher acquaints the students with the problem to be investigated and the reasons for investigating it. Most of this information should be obtained from the guided discussion. After the completion of all laboratory work, the students gather for a postlaboratory session in which the students correlate their data and draw their conclusions. The teacher's guide for the laboratory manual is undoubtedly the most comprehensive manual ever developed for teacher assistance. It advises of the problems that the students are likely to encounter and provides sets of sample student data. In addition, it gives the philosophy of each experiment and goes into great detail about what facts and ideas should be brought out in the prelaboratory and postlaboratory discussions. The teacher's guide for the text provides instruction in the presentation of the concepts and the background information for answering the anticipated student questions.

The developmental sequence for this course was brief. The initial version was prepared during the summer of 1959 and tested that fall. A second revision was made the following summer with a larger group of students being used for the trial period the following year. A third revision followed in 1961. Each trial required extensive revisions with the final copy being completed in late 1963. The complete set of materials was offered for general public use in 1964. The extent of the required revisions and the limited time in which they were completed produced a final version that contained several mistakes. Ideally another revision would have been forthcoming but to this date none has appeared.

The quality of the mechanics of the published materials is excellent. The pictures, charts, and graphs used in the text require careful consideration by the teacher and student. The short biographical sketches of famous scientists introduce a flavor of history that enhances the main theme without complicating it.

The following materials are available for this course:

1. Chemical Bond Approach Project, *Investigating Chemical Systems* (New York: McGraw-Hill Book Company, Inc., 1963).
2. Chemical Bond Approach Project, *Investigating Chemical Systems Teachers Guide* (New York: McGraw-Hill Book Company, Inc., 1964).
3. Chemical Bond Approach Project, *Chemical Systems* (New York: McGraw-Hill Book Company, Inc., 1964), pp. 752.
4. Chemical Bond Approach Project, *Chemical Systems Teacher's Guide* (New York: McGraw-Hill Book Company, Inc., 1964).

The Chemical Bond Approach Project's efforts have not been widely adopted. Initial data from the field indicated that it was successful when average or above average students were instructed by superior teachers. The degree of sophistication in the approach and the emphasis that was placed on conceptual models undoubtedly was rejected by the majority of the high school chemistry teachers as being unrealistic. The Project did not make a sustained effort to develop audiovisual materials or other supplementary materials for the course. The other high school chemistry curriculum project, CHEMS, made this effort and the teachers adopted the program with the largest variety of material, references, and supplements.

REFERENCES

Ashmore, P. G., "On Teaching High School Chemistry," *Science*, 46(6):1312-1314 (1965).

Montean, J. J., R. C. Cope, and R. Williams, "An Evaluation of CBA Chemistry for High School Students," *Science Education*, 47(1):35-43 (1965).

Pode, J. S. F., "CBA and CHEM Study: An Appreciation," *Journal of Chemical Education*, 43(2):98-103 (1966).

Strong, L. E., and Wilson, M. K., "Chemical Bonds: A Central Theme for High School Chemistry," *Journal of Chemical Education*, 35(2):56-58 (1958).

Walker, Noojin, "CHEM Study, CBA, and Modern Chemistry: A Comparison," *School Science and Mathematics*, 67(7):603-609 (1967).

Chemical Education Materials Study

The Chemical Education Materials Study program is an outgrowth of a committee established by the American Chemical Society in 1959. The charge to the committee was to examine the current high school chemistry courses and determine what needed to be done to improve them. The recommendations of this committee were the basis for a grant from the National Science Foundation to prepare and field test materials that would improve the instruction of high school chemistry. This committee was composed of high school and college chemistry teachers as well as industry representatives. The conclusion of the committee was that the high school student should be more involved with the investigation processes of chemistry rather than merely memorizing large quantities of facts. Principles relating observations and other factual information were to be emphasized. The committee assumed that a student trained in this manner would thus be better equipped to continue his studies in the college general chemistry course. The committee agreed that a more interesting and understandable course would attract more students to chemistry as a profession.

The objectives of this group were similar to those developed for the Chemical Bond Approach Project. The emphasis on individual objectives and the strategy for implementing those objectives were different. Among those objectives considered important by the developers are:

1. To develop the relations between experimental data and theory in a systematic fashion.
2. To build a conceptual model of chemistry that rests on the principles underlying molecular structure, energy, and combination.
3. To learn the principles of chemistry by application and use rather than by memorization.
4. To construct a sequence of instruction that would encourage the students to pursue science as a career.

Fourteen chemistry teachers from both the high school and college level prepared the initial version of the program in 1960 with the first trial coming in the 1961-1962 school year. Although all involved made significant contributions, two are given primary credit for the book in its final version: Professors G. C. Pimentel and J. A. Campbell.

As the writing progressed, it became apparent that only a selected amount of material could be included in the final text. As a result, specific criteria were developed to determine which principles were to be included. Basically these criteria are as follows:

1. Is the principle a necessary prerequisite to future material?
2. Is the principle comprehensible to the average high school student?
3. Does an experiment exist that will demonstrate that principle in the high school chemistry laboratory?
4. Can the principle be related to the other principles presented in the text so it may be reinforced?

The principles and ideas of chemistry were to be discovered by the student from carefully selected laboratory experiments. To accomplish the stated objectives the book was constructed in three sections. The first section has a heavy experimental emphasis with great detail in directions, to insure that the student had all of the relevant experiences before the topic was discussed. This section serves as an overview of the course. The second section stressed the important chemical principles that could be determined from experience and films that had been prepared by the CHEMS group. The last section includes descriptive material designed to allow the student to apply the principles and ideas to current problems such as biochemistry and astrochemistry.

The materials produced for this program follow a good developmental sequence. The initial version was used by a small group of teachers the first year; these teachers in turn assisted in the first revision in the summer of 1961. The test group was enlarged for the next school year, and the feedback was used to prepare a second revision in the summer of 1962. The final version was published in 1963 by the W. H. Freeman Publishing Company as *Chemistry: An Experimental Approach.* The revisions required for this material were not as extensive as for the CBA book, and, as a result, fewer mistakes were made in the initial hard cover versions. Tests were developed and produced which measured the student's achievement of these new and different objectives. A teacher's guide was also prepared from the suggestions of the initial teachers' group which used the materials; this teacher's guide provided needed background information in various areas. Films and film loops were also produced with the goal of making the entire CHEMS program a cohesive unit.

The CHEMS program was immediately more successful than the CBA. Within two years approximately 15 percent of the high school chemistry students in the United States were using CHEMS material. This success is correlative to the CHEMS group's attempt to write the material for student success. Another reason was the more extensive supplemental materials produced by the CHEMS program. Both books were well written but the CHEMS book was more understandable to the student than was the CBA book. With the initial success and the objectives that were accomplished, three separate revisions were authorized by the Chemical Education Material Study committee. With the completion of these revisions the committee disbanded and is no longer controlling these materials.

The revisions which were authorized are:

1. Cotton, F. A., and L. D. Lynch, *Chemistry: An Investigative Approach* (Boston: Houghton Mifflin Company, 1968).

2. O'Conner, P. R., J. E. Davis, E. L. Haenisch, W. K. MacNab, and A. L. McClellan, *Chemistry: Experiments and Principles* (Lexington, Mass.: D. C. Heath & Company, 1968).
3. Parry, R. W., L. E. Steiner, R. L. Tellefsen, and P. M. Dietz, *Chemistry: Experimental Foundations* (Englewood Cliffs, N.J.: Prentice-Hall, Inc., 1970).

The Cotton and Lynch version is closest to the original version of CHEMS because it was updated with the feedback from the experiences of many teachers in several schools. This is the only version that retains the experiments in the book itself to indicate the relative placement with respect to the principles that they demonstrate. A teacher's handbook is available that is extremely helpful to the teacher; it provides answers to questions, background information, and information about the experiments. This version does not present an adequate quantum mechanical concept for the student. The original and the new supplemental audiovisual CHEMS materials are available from other sources such as Modern Learning Aids.

The revision by O'Conner, et al., separates the experiments from the text into a laboratory manual and includes more experiments than the original version. It also provides a teacher's handbook to assist the teacher in planning and organizing the program. The areas that were deleted from the original version by CHEMS are also not presented here. Quantum mechanics is barely mentioned and molecular orbitals are totally neglected. Two of the five authors were part of the original writing committee and helped maintain the perspective of the original objectives. The objective of O'Conner, et al., is to provide a more descriptive text than the original version.

The Parry, et al., revision also separates the experiments into a laboratory manual. A teacher's guide is provided with the text. This revision is more fully descriptive than the Cotton and Lynch or O'Conner, et al., revisions, and has added material that is not included in the others because of its later printing date.

All three versions maintain the integrity of the original work. The selection among these books will be a matter of the teacher's own personal preference. All books are well illustrated and seek to emphasize model building by the students. The primary difference among them is the quantity of additional descriptive material and the value to be derived from the addition.

Films and film loops are available from Modern Learning Aids. The films allow the teacher to show events that would not normally be available for student observation. The film loops help to augment the laboratory program by reducing the laboratory time of the students and the teacher. The film loops should play a significant role in the use of any one of the three revised texts.

REFERENCES

Ashmore, P. G., "On Teaching High School Chemistry," *Science*, 46(6):1312-1314 (1965).

Campbell, J. A., "The Chemical Education Materials Study," *Journal of Chemical Education*, 39(1):2-4 (1961).

Pode, J. S. F., "CBA and CHEM Study: An Appreciation," *Journal of Chemical Education*, 43(2):98-103 (1966).

Project Physics

The Harvard Project Physics group was organized in 1963 in response to concern about the status of physics in American high schools. The project directors are Gerald Holton, F. James Rutherford, and Fletcher Watson. The objective of this group was to produce a physics course that would present aspects of physics that were not emphasized in existing high school physics courses, provide the teacher with instructional and media assistance, provide a complete set of materials for effective instruction, and appeal to a greater proportion of students than are presently enrolled in high school physics courses (Holton, 1967). This increase in diversity in physics course offerings was viewed as an important means of increasing enrollments in physics courses since about 80 percent of the high school students had been electing *not* to take any of the existing physics courses. However, the Project Physics course was not intended as a replacement or a substitute for the PSSC course (Holton, 1967; Watson, 1967). Whether both Project Physics and the PSSC course will survive in a symbiotic arrangement remains to be seen.

The Project Physics course originated with Harvard Project Physics, a curriculum improvement group organized at Harvard University with financial support from both federal agencies and private foundations. Project Physics, Inc., is in charge of the production of this course and the associated materials. Holt, Rinehart and Winston, Inc., is the official distributor for all of the Project Physics printed materials, and the laboratory equipment can also be obtained from them.

Project Physics used the services of educational and media specialists to a greater extent than any of the previous curriculum development projects. Considerable effort was placed in the development of effective instructional materials that were also visually attractive and interesting. The evaluation of the course during its developmental period was exceptionally comprehensive.

A complete package of instructional materials has been produced by the Project Physics group. The textbook, or student guide, is divided into six units:

1. Concepts of Motion
2. Motion in the Heavens
3. The Triumph of Mechanics
4. Light and Electromagnetism
5. Models of the Atom
6. The Nucleus

The interrelationship of these units and their concepts is exceptionally well done. The emphasis is upon those topics of physics that are significant and important. Considerable attention is also given to both the history and philosophy of science and the interaction between science and society. This seems to be one of the most striking differences between Project Physics and other courses or texts. The human aspects of science and scientists are also illustrated with short quotations. The illustrations in the text are excellent and add much to the attractive appearance of the text. Some illustrations can be used as sources of data. Throughout the chapters brief questions appear that are related to the previous material. These questions are intended for use as a self-check device for the students, and answers are provided in the back of the text. A set of problems is included at the end of each chapter. Some of the problems are intended for use as topics of class discussion and others are intended for individual practice in the application of the concepts.

The Project Physics *Readers* are diverse collections of articles and excerpts from books and other sources that are related to the units in the textbook. The works of Galileo, Copernicus, and Lord Kelvin are represented along with a chapter from Hoyle's science fiction novel *The Black Cloud*, and articles relating physics to both music and sports. These readers are neither textbooks nor library books, but are rather supplementary materials that are correlated with the course, and are of a diverse nature so that considerable differences in student interest can be accommodated.

The *Student Handbook* serves as a laboratory guide and provides direction for student experiments. The laboratory program of Project Physics is an important aspect of the course instruction and concept development for the students. Different approaches to laboratory investigations in a given area are often being used by different groups of students at the same time, with a postlab group discussion being used to compare and relate results. This multiple approach to instruction is used frequently and effectively in the Project Physics Course.

About fifty film loops have been developed for the Project Physics course. Some of these are intended for use as sources of data in laboratory situations or for independent study. Other film loops are intended as demonstrations. Three 16mm sound films have been produced, but for the most part existing films, especially those from the PSSC course, have been incorporated into the course of study. A series of overhead transparencies has also been produced which are related to specific sections of the text. Some programmed instructional materials have been developed. These programs are brief, dealing with only a specific concept or skill. Many of these materials, which have been developed for the Project Physics course, can be effectively used by teachers who do not use the course.

The *Project Physics Teacher Resource Book* is designed for use with this course, but it too can be valuable for teachers who are not using the Project Physics course. Along with an extensively annotated version of the textbook, the *Teacher Resource Book* includes descriptions of the laboratory experiments, suggested demonstrations, notes on films and reader articles, bibliographies, and a detailed multimedia schedule for presentation of the instructional materials. Project Physics has also developed its own achievement tests. These are reviewed in the section on tests.

Project Physics is an extremely attractive course that emphasizes good physics, both as it is recognized by physicists and in the sense of providing the student with powerful instructional material. It should be considered by schools and physics teachers who are contemplating a change in their existing physics program, and also examined carefully for materials or ideas that can be placed into use in other physics programs. Further information on Project Physics can be obtained from the *Physics Teacher* issue from which the references (Holton, 1967; Watson, 1967) were taken. A concise description of the Project Physics course appears in the *Project Physics Teacher Resource Book* for unit 1. The Project Physics course has also been described by Hurd (1969).

REFERENCES
Holton, Gerald, "Project Physics: A Report on Its Aims and Current Status," *The Physics Teacher*, 5:199–211 (May 1967).
Holton, Gerald, F., James Rutherford, and Fletcher C. Watson, *The Project Physics Course Teacher Resource Book* (New York: Holt, Rinehart and Winston, Inc., 1970).
Hurd, Paul DeHart, *New Directions in Teaching Secondary School Science* (Chicago: Rand McNally and Company, 1969).
Watson, Fletcher, "Why Do We Need More Physics Courses?" *The Physics Teacher*, 5:212–214 (May 1967).

PSSC Physics

The acronym PSSC is derived from Physical Science Study Committee, the group that developed this course. Work on this course began in 1956. It was the first course developed by a committee of scientists, high school teachers, and specialists in various fields of science education, and also the first course development program on a truly national basis. Substantial funding for this group's activities was provided by foundation grants and grants from the federal government agencies, especially the National Science Foundation. Educational Services, Inc., is currently in charge of the revision and development of the PSSC course. Raytheon Education Corp., through its subsidiaries, Heath Publishing and Macalester Scientific Co., is the textbook publisher and primary distributor of PSSC laboratory materials.

The PSSC course development received much support during the post-Sputnik period of intense public concern about the quality of science instruction in the United States. This course was designed for the students who were electing to take physics in high school. The PSSC course was placed into extensive use; estimates of the enrollments in courses using PSSC materials as the principal means of instruction have been as large as 40 to 60 percent of the total high school physics enrollments (Haber-Schaim, 1968; Hoffart, 1967).

During the period of the development and implementation of the PSSC physics course, the percentage of high school students enrolled in physics courses declined substantially (Ellis, 1967). Haber-Schaim (1967) contends that increased enrollments were not really the objective of the PSSC course, but it has also been suggested that the extensive adoption of the PSSC course and the difficulty of this course were important factors in this decline in enrollments (Jacob, 1968).

The authors of the PSSC physics course intended to present physics in a manner that was consistent with the physicist's interpretation of his discipline. The course contains significant departures from both the content and the instructional procedures of the pre-1960 high school physics courses.

In the textbook, topics such as simple machines, fluid mechanics, and calorimetry were omitted. This elimination of many topics that were traditionally found in physics textbooks was accompanied by a more thorough coverage of the remaining topics. Mechanics and wave phenomena were given special emphasis. Processes such as measurement, model building, and data analysis were also emphasized. The textbook is well written and presents good physics as defined by many physicists, but neither of these aspects of the book is likely to be appreciated by the student unless he is a good reader who is capable of formal, abstract reasoning. The illustrations are excellent and include much effective use of stroboscopic photography. Photographs from the PSSC text can now be found in almost all introductory physics textbooks. At the end of each chapter, there is a set of problems for "home, desk, and lab." These exercises require careful thought and are an intrinsic part of the PSSC course. In the second edition, some problems are identified as referring to a specific section of the chapter to provide a means for the student to check his understanding of the important concepts of the chapter. References to supplementary reading materials are also included.

Laboratory work plays an important role in the PSSC course. The developers of this course not only designed experiments but also designed a completely new set of apparatus. Although much of this equipment was of unique design, the most significant departure from earlier equipment was the emphasis given to its simplicity and low cost. High schools could thus have enough equipment for students to work individually or in small groups. A student laboratory guide is provided for the students.

The PSSC films represent a major contribution to physics education. These fifty-five films cover most aspects of the PSSC course; they are available from Modern Learning Aids. Many of these films can be used effectively outside of the PSSC course as well; the *Frames of Reference* film by Hume and Ivey is an excellent example.

In terms of the reading level of its textbook and the sophisticated reasoning that is expected, the PSSC course is aimed at the students of above average academic ability. It can be used successfully by these students, and especially by those students who intend to major in science or mathematics in college. However, the course need not be restricted to these students. Some teachers have been able to modify the course so that reading skills are deemphasized and less abstract versions of the lessons are presented. These teachers have had success with students with lower academic ability (Matthews, 1967); however, this use of the PSSC course is exceptional.

Descriptive reviews of the PSSC textbook and course have appeared in Hurd (1969) and Nedelsky (1965). The tests of the PSSC course are reviewed in the section on testing.

REFERENCES

Ellis, Susanne D., "Enrollment Trends," *Physics Today*, 20(3):75-79 (March 1967).

Haber-Schaim, Uri, "The PSSC Course," *Physics Today*, 20(3):26-31 (March 1967).

Haber-Schaim, Uri, "The Use of the PSSC Course in the United States," *The Physics Teacher*, 6:66-67 (February 1968).

Hoffart, Ervin H., "PSSC," *The Physics Teacher*, 5:35 (January 1967).

Hurt, Paul DeHart, *New Directions in Teaching Secondary School Science* (Chicago: Rand McNally and Company, 1969).

Jacobs, Charles, "Physics—Phooey," *The Physics Teacher*, 6:450-453 (December 1968).
Matthews, Richard E., "PSSC and the Low Ability Student," *The Physics Teacher*, 5:34-35 (January 1967).
Nedelsky, Leo, *Science Teaching and Testing* (New York: Harcourt, Brace Jovanovich, Inc., 1965).

Engineering Concepts Curriculum Project

The Engineering Concepts Curriculum Project, or ECCP, is another development that originated in the early 1960's with the belief that a greater diversity of course offerings in physics and physical science would be an effective means of combatting declining enrollments in these areas. Work on this course was begun in 1963, with trial class testing beginning in 1965. After several summer writing conferences, commercial versions of the ECCP text, *The Man-Made World*, and laboratory materials were produced. The publisher of the ECCP material is McGraw-Hill and the apparatus is available from several suppliers whose addresses are given in the article by Liao and Piel (1970). The Engineering Concepts Curriculum Project headquarters are located at the College of Engineering, State University of New York, Stony Brook, New York 11790. By the academic year 1969-1970, it was estimated that 7,000 students in 200 schools across the country were enrolled in ECCP courses.

ECCP is neither a physics course nor a preengineering course, but is intended to be a laboratory science course that is independent of existing science courses. Consequently, it has no science prerequisites, although two years of mathematics are required. ECCP should be appropriate for the average college-bound student, including students who do not intend to major in science, engineering, or mathematics. The course is designed to emphasize technology and its implications for its creator, man. Instead of starting with fundamentals and culminating in applications as most physics courses attempt to do, ECCP begins with familiar, observable problems and then proceeds to basic principles. Many current problems such as population growth, traffic flow, waste disposal, and optimal yield are discussed, and important aspects of technology, including computers and logic circuits, modeling processes, and the general problems of control and feedback systems are emphasized. A special laboratory equipment package is required that includes a logic-circuit board, a simple analog computer, and some electronic equipment (David and Truxal, 1967; Liao and Piel, 1970).

ECCP should not be considered as a substitute or a replacement for a physics course, particularly since the physics content that is involved in ECCP is incidental. However, ECCP should be considered as a laboratory science course as an alternative to a course in physics, especially for those students who do not normally elect to take physics courses. ECCP courses would be especially attractive if time were available for student use of a computer; this can sometimes be arranged with the school district's own computer or a machine in a local business or industry. It should also be noted that ECCP and physics courses need not be mutually exclusive; indeed, the classes of one could be fertile recruiting territory for the other.

REFERENCES
David, Edward E., and John G. Truxal, "Engineering Concepts," *Physics Today*, 20(3):34-40 (March 1967).
Liao, T., and E. J. Piel, "Let's Get Relevant," *The Physics Teacher*, 8:74-82 (February 1970).

Other Chemistry Textbooks

The CBA and CHEMS programs have had their effect on the other chemistry texts that are now available. The books currently being adopted stress the principles more than the descriptive behavior of chemistry. The total number of chemistry books presently available does not permit a detailed analysis. A partial listing of some of the more recent books that have received a moderate amount of success is given. Those texts in second or third revisions will probably be available for selection but you should check to see if newer editions are now available.

In 1972 the first of a new wave of supplemental books and textbooks was offered to the high school and college chemistry teacher. These books examined the principles of chemistry from an environmental perspective. In an attempt to make the concepts relevant, this context was chosen because of the public interest. One benefit of this emphasis will be a recognition of the basically interdisciplinary nature of the problem. This may break down some of the walls that have been used to compartmentalize the sciences. The teacher should watch the periodicals for advance notices and have his name placed on several publishers' mailing lists.

Some of the chemistry books presently available are:

1. Bickel, C. L., et al., *Chemistry: Patterns and Properties* (New York: American Book Company, Litton Educational Publishing Inc., 1971).
2. Choppin, G. R., et al., *Chemistry* (Morristown, N.J.: Silver Burdett Company, 1970).

3. Eblin, L. P., *Elements of Chemistry*, 2d ed. (New York: Harcourt Brace Jovanovich, Inc., 1970).
4. Greenstone, A. W., *Concepts in Chemistry*, 2d ed. (New York: Harcourt Brace Jovanovich, Inc., 1970).
5. Metcalfe, H. C., *Modern Chemistry*, 4th ed., (New York: Holt, Rinehart and Winston, Inc., 1970).
6. Smoot, R. C., et al., *Chemistry: A Modern Course*, 3rd ed. (Columbus, Ohio: Charles E. Merrill Publishers, 1971).
7. Toon, E. R., et al., *Foundations of Chemistry* (New York: Holt, Rinehart, and Winston, Inc., 1968).

All of these textbooks that were designed for use in high school chemistry courses also have an extensive amount of supplementary material available. The teacher will have a choice of a laboratory manual, a teacher's manual, a set of tests developed specifically for use with the text, and supplementary readings for the students. Many publishers of textbooks are also producing a wide variety of audiovisual materials, particularly film loops, slides, and transparencies. The teacher will need to contact the publishers directly and determine what supplemental aids for specific texts are available and evaluate these aids according to the previously discussed criteria for use in the selection of textbooks.

Other Physics Textbooks

Genzer, Irwin, and Phillip Younger, *Physics* (Morristown, N.J.: Silver Burdett Company, 1969), pp. 762 + v.

In their preface the authors state that this book was produced in an attempt to satisfy the need for a program that would combine some ideas from the PSSC physics course with other innovations in teaching. The influence of the PSSC course is evident, and the writing style seems similar to that of the PSSC text. Specific changes from the PSSC program are the inclusion of topics such as calorimetry and a reduction in the emphasis given to mechanics. The book is quite comprehensive in its coverage. In addition to the material normally found in high school physics texts, a unit on astronomy has been included. Graphical analysis is emphasized in the study of motion, and delta notation is introduced early and is used extensively. Both English and metric units are used in examples and problems.

This book is visually attractive and includes some useful pedagogical features. "Do You Really Understand?" questions are included in the chapters for use as a self-check on the concepts in the chapter. These questions include simple problems and applications as well as questions involving recall of earlier material. "Try it Yourself Activities" are included for independent investigations by the students either at school or at home; only simple materials are required for these activities. Questions and problems are included at the end of each chapter. Some of the questions and problems are intended to provide practice with application of the concepts, whereas others are more challenging. The illustrations in the text are well done and attractive. A unique feature of this text is the "picture essay," brief, extensively illustrated studies devoted to measurement, motion, energy, and color. Some of these could be used effectively even by teachers who do not use this text regularly.

Relatively few specific criticisms can be leveled at this book. The chapters on motion deal with speed and acceleration; however, at the beginning of the chapter on force, velocity is introduced in a warning that cautions the student not to confuse speed and velocity. The book involves a considerable amount of reading, and many lengthy discussions are provided at a reading level that seems comparable to the PSSC text.

This text presents a thorough treatment of physics. The teacher who has had success with PSSC, but prefers a more traditional approach, would do well to consider this text. A teacher's edition of the text, tests on Ditto masters, and an accompanying laboratory manual are also available.

Lehrman, Robert L., and Clifford Swartz, *Foundations of Physics* 2d ed. (New York: Holt, Rinehart and Winston, Inc., 1969), pp. 757 + x.

In the preface the authors express the philosophy that the fundamentals or foundations of physics are intimately related to modern physics research, and it is these foundations that should be studied in an introductory course. However, it is the authors' contention that these fundamentals should be studied with an emphasis upon their operational nature instead of as a set of overly concise textbook definitions. This book is consistent with that philosophy.

The text provides a rigorous treatment of physics, without the use of a glossary or similar emphasis upon terminology but with considerable emphasis upon mathematics. The mathematical tools are introduced in the first two chapters, along with a discussion of the operational nature of physics, measurement, and the units of length, mass, and time. The study of motion follows this introduction, and graphs and delta notation are used effectively in the chapters on motion. The figures and the questions in the chapters are important and must be studied carefully and thoroughly. In the preface the authors use energy as an example of a concept that is inadequately described by a simple definition such as the "ability to do work," and their treatment of energy in Chapter 8 exemplifies this philosophy. Although simple machines have been removed from most high school texts, they are used here to illustrate and introduce work and energy. This discussion is clear and well done. Heat energy and wave motion each receive two chapters, whereas electricity and magnetism is covered in five chapters, and four chapters at the end of the book are devoted to atomic and nuclear physics. In addition to these topics, which are normally associated with a high school physics course, there is a chapter (Chapter 7) devoted to relativity and another chapter (Chapter 23) devoted to astronomy.

Care has been taken in this text to make it pedagogically useful. Questions are included within each chapter, and a list of answers is provided at the end of the chapter. These questions should be studied carefully by the student as he goes through the chapter, since they serve as a self-check upon understanding, and to assist in the further development of the concepts presented in the chapter. An extensive set of problems is included at the end of each chapter, which are generally more difficult than those included as questions within the chapters. Although most problems and examples use metric units, some problems also use English units. A list of suggested readings from other books, paperbacks, or magazines such as *Scientific American* is provided for each chapter. The layout of the book, with its wide margins and black and green printing, is attractive. This two-color format is effectively utilized in the graphs and line drawings. One-page biographies of physicists are included in the chapters that are related to their work.

This book presents a rigorous treatment of physics, and is appropriate for students who are at ease with mathematics and who are good readers. It can be recommended to the teacher whose students fit this description, who agrees with the authors' philosophy regarding the teaching of physics, and who can select his course material, since the text contains more than enough for a one-year course. A laboratory manual, with open-ended experiments related to this text, and a teacher's edition, including suggested tests, are both available.

A review of this book appeared in *The Physics Teacher* of November, 1970.

Stollberg, Robert, and Faith Fitch Hill, *Physics Fundamentals and Frontiers* (Boston: Houghton Mifflin Company, 1965), pp. 695 + vii.

The material in this book has been selected and arranged so that the book can be used with or without an emphasis upon a single, unifying theme. However, a core of basic principles is intended for mastery by all students, and these portions of the text have been identified in the teacher's guide edition. Material to supplement this basic core can be selected from a wide range of topics that have been effectively integrated into the text.

The appearance of the book is attractive. The extensive use of color provides emphasis and makes the diagrams more effective. In general, the illustrations are pertinent and useful. Considerable care has been taken to make the book useful for self-study. Within each chapter, the material has been divided into topics that would roughly correspond to daily lessons. These topics are followed by Self-Check questions, Discussion Questions, and Problems. The Self-Check questions are intended to provide an opportunity for the student to see if he can recall the important concepts in the material he has just read. The Discussion Questions are designed to encourage the student to apply text material to his previous experience. Many of these questions are excellent and will require serious thought by the student. They could be used as a basis for class discussion as well as for evaluation of the students' understanding and ability to apply the concepts in new situations. Worked examples are included; the solutions are very clearly shown and are arranged to illustrate the proper use of significant figures and dimensional analysis. Problems are included for the quantitative application of the concepts. These problems have been divided into A groups, which are intended for all students, and B groups, which are more difficult problems and derivations intended to challenge the more capable students. Both metric and English units are used in the text and in the problems.

The appendix includes tables of trigonometric and logarithmic functions, a list of important equations, and values of physical constants. It also includes a brief mathematical review. Algebraic and mathematical operations are illustrated using equations from physics, and a few practice problems are also included. The index is complete and will increase the value of the book as a reference.

This book seems suited to a good, thorough presentation of physics. Some of its shortcomings are the brief, ten-page treatment devoted to accelerated motion, the failure to continue to effectively use the delta notation after it has been defined, and the failure to label vector quantities, e.g., electric field intensity, in their defining equations, although this may be made clear in the subsequent text material. The use of both metric and English units may be an advantage or a decided disadvantage of this book, depending upon the teacher's point of view. The reading style of this book is smooth and interesting for average or good readers, but adjustments such as greater emphasis upon the self-study questions might be necessary for poor readers. After considering the abilities and interests of his students, adjusting the rate of coverage of the basic material and the amount of supplementary material that is included should enable the teacher to use this book effectively.

Another review of this book has appeared in *The Physics Teacher*, 4:89-90 (February 1966).

Taffel, Alexander, *Physics: Its Methods and Meanings* (Boston: Allyn & Bacon, 1965), pp. 566 + ix.

The theme of this book is the study of matter and energy. In the introduction these two concepts are defined in a cursory manner, but these definitions are expanded in the rest of the book as the theme is developed. Each chapter is related to this theme.

The text is divided into seven units:

1. Measurement
2. Force, Motion, and Energy
3. Heat and the Structure of Matter
4. Wave Motion and Light
5. Electricity
6. Electromagnetism
7. Quantum Theory and Nuclear Physics

It is the author's intention that all of these units should be included in a one-year course.

The pedagogical features of this book are conventional. A brief summary is included at the end of each chapter. The questions included are divided into two groups; Group 1 consists of straightforward, essentially recall questions intended to allow the students to test their understanding of the basic concepts. The Group 2 questions are more challenging and are intended to provide opportunities to relate the concepts to physical situations. Problems are included with each chapter and are similarly divided into two groups according to their difficulty. Student activities that can be done at home with simple materials are also described, and references for supplementary reading are included. Both English and metric units are used in the book. In the discussion, almost all units from the pound mass to the dyne are mentioned, but only absolute units of force and mass are used in the examples and problems. Most of the illustrations are well done, and the red and black printing in the line drawings is effective.

This book does have some undesirable aspects. Significant figures are discussed, but the answers for problems and examples are not always consistent with their use. The contradiction between the stated rules for significant figures and the results of the examples and problems may be confusing to the student. Figure 30-1 is inaccurate since it shows alpha and beta particles undergoing deflections of equal magnitude in an electric field. The definition of the second is given in terms of the year 1900. Celsius is given as an alternate name for the Centigrade scale and is not used further.

For the teacher who wants a textbook with a comprehensive treatment of physics with an emphasis upon modern physics, who wants to use both metric and English units, and who does not find the undesirable aspects of this book to be totally unacceptable, this text would be adequate. A teacher's manual is available that divides the book into 100 single-lesson segments and also provides answers to the questions and solutions for the problems. A laboratory manual is also available.

A brief review of this book appeared in *The Physics Teacher* of February, 1966, on page 89.

Williams, John E., H. Clark Williams, Frederick E. Trinklein, and Ralph W. Lefler, *Modern Physics* (New York: Holt, Rinehart and Winston, Inc., 1968), pp. 707 + xi.

This book is the successor to *Modern Physics* by Dull, et al., and the influence of the earlier editions is evident. An extensive amount of material is included in this book; the index contains over 4,000 entries ranging from AC to zinc electroplating. A policy of careful selection among these topics will have to be practiced if this book is used. Nearly all of the traditional topics of a high school physics course are included in a conventional order except for nuclear energy, which follows a standard treatment of energy. A unit on the frontiers of physics has been spliced onto the end of the book.

Each chapter is divided into sections that are followed by questions relating to the previous material. Very few of these questions involve anything more than recall of the material. Problems are also given to provide quantitative practice with the concepts and relationships. Both the questions and the problems are divided into sets labeled A and B. The B questions and problems are more difficult, but since some of them are good, thought-provoking questions it does not seem a good procedure to relegate them to a supplementary or optional role.

All units are consistent with 1970 definitions with the exception of the coulomb and the ampere. Apparently for pedagogical reasons, the coulomb is defined in terms of the charge on an electron and the ampere is defined as a current of one coulomb/second. Only MKS units are used in this text and its problems.

The book contains many illustrations, almost one figure per page. The line drawings are well done, but some of the photographs are of doubtful value such as the picture of a balance, Figure 2.5, or the picture of a sonometer, Figure 14.8.

At the end of the book are a mathematics refresher, appendices, and a glossary. The mathematics section is essentially limited to calculation procedures. Neither a set of practice problems nor a review of algebra is included in this section. Appendix A consists of a list of formulas, some of which are of doubtful value for a high school physics course (e.g., the transconductance of a vacuum tube) or are presented in a questionable form. The equation for angular velocity is listed here as $\omega = \theta/t$ despite the use of the more desirable delta notation in Chapter 5. Appendix B is a useful collection of data tables, values of physical constants, and trigonometric and logarithmic tables.

This book contains enough materials for a very complete physics course. Its primary disadvantage would seem to be an excessive concern with terminology. The format of the review outlines at the end of each chapter as well as the questions that emphasize recall would seem to encourage memorization of the appropriate passages from the text or glossary. A teacher who can get his students to avoid this type of rote learning and who can lead his classes through selected portions of this book should be able to have his classes achieve a mastery of conventional, although not necessarily *modern* physics.

This book was reviewed by N. C. Little in *The Physics Teacher*, 7:295 (May, 1969).

Standardized Tests

To evaluate student learning, the teacher has the choice of constructing a test or using a standardized test. The teacher-constructed test will probably reflect a more accurate picture of the actual instruction that has taken place in the classroom. The teacher-constructed test has a number of limitations that are discussed at length in any standard text on tests and measurements. The teacher or his administration will frequently want to know how student achievement in their school compares to national achievement levels. For this reason, the teacher will have to select one or other of the currently available standardized tests.

The teacher must be aware that even standardized tests are not the same. With the advent of the curriculum projects, the nation's students were receiving two entirely different types of instruction. The existing tests were biased toward the ability to memorize large amounts of factual information and could not be used to evaluate student achievement under the new curricula that emphasized principle learning and problem solving. The curriculum groups prepared and standardized their own tests to evaluate their objectives. When selecting a standardized test, the teacher must choose one that reflects the type of curriculum that the student studied. If it was factually oriented, then the test used should be factually oriented and conversely, the principle oriented test should be used for the curriculum emphasizing principles. No one test can do an adequate job of testing both types of curricula.

Because of these differences, the interpretation of the test scores must be made with great care. It should serve more as an aid to an evaluation of instruction than of student achievement. Teachers have tried to combine the best of both curricula into a single course, which may produce test scores based on fact and principles below the average for either test. The teacher(s) instructing in a given school should examine the test to determine if the success or failure of the students with respect to the standardized test used can be attributed to the deviation of the classroom instruction from the norm of each test. This can not be overemphasized if only one standardized test will be used to measure student achievement at the end of the year.

Each textbook will have a set of tests that are available for purchase for use in measuring achievement at regular intervals. The curriculum projects developed a series of tests for use throughout the year. If the teacher or school system desires to measure achievement against the national norms, it should be done on a regular basis.

The teacher should also secure information that will aid him in the interpretation of scores from the publishers of the tests. There is a definite need to examine the credentials of those involved in the preparation of tests, the characteristics of the group of students on which the test was normed, the time of year when the test was given to the sample group, and if the test was rewritten and tested again. The score on one final standardized achievement test should be translated into a grade for the student only if the teacher is satisfied that the student had a reasonable probability of success.

Before selecting a standardized test, the teacher should select samples from the sources listed and expose those tests to an analysis such as that performed on the following two tests.

Project Physics Tests, distributed by Holt, Rinehart and Winston, Inc., 1970. There are six tests in the Project Physics series, one for each of the six basic units. Four forms are available for each test. Forms A and B are parallel; each has fifteen multiple-choice questions and six to eight problem-and-essay questions. The problem-and-essay questions are divided into two sets, and the student is to answer a specified proportion of the questions in each set. Allowing the student to select the questions to be answered reduces the importance of specific information, but it also reduces the reliability of the test scores (Nedelsky, 1965). However, this opportunity for choice is popular with the students. Form C consists of forty multiple-choice questions. Form D consists of eight problem-and-essay questions; four out of five questions are to be answered in group one and two out of three questions are to be answered in group two. Numerical values for physical constants and the equations that may be useful on the test are provided in the test booklet. Forty minutes working time is required for each test, and suggestions on how the student should distribute his time among the questions are provided.

Answers for the multiple-choice questions are provided in the *Teacher Resource Book* and the relevant section in the text is indicated for each question. Outlines of the answers for the problem-and-essay questions are also included.

The entire Project Physics course was extensively evaluated during its development. This is reflected in the test items, answers that are accurate and multiple choice items that are unambiguous. Despite this extensive evaluation program, no data of reliability is supplied with the tests. The *Teacher Resource Book* provides the proportion of students correctly answering the multiple choice items, but this data is not given for all tests and there is no information about the sample size from which this data was obtained. The values given are for individual items and must thus be used and interpreted cautiously. The proportion of students correctly answering the item is not always reproduced when the same items appear on different forms:

Table 4-1. Largest Discrepancies in Proportion of Test Sample Answering the Same Item Correctly on Different Forms. [Data taken from Holton, Gerald, F. James Rutherford, and Fletcher C. Watson, *The Project Physics Course Teacher Resource Book* (New York: Holt, Rinehart and Winston, Inc., 1970)].

Test	Form	Item	Proportion of Test Sample Answering Item Correctly	Form	Item	Proportion of Test Sample Answering Item Correctly
1	A	5	.944	B	7	.500
2	A	11	.67	C	8	.78
3	A	9	.55	C	24	.95
4	B	1	.34	C	11	.59
5			Information Not Available in *Teacher Resource Book*			
6						

The test authors have attempted to construct the tests so that the mean grade would be about 70 per cent (Winter, 1967). Many of the questions on the tests seem to be in the lower categories of Bloom's taxonomy (Bloom, 1956). The essay tests do provide some opportunities to encourage and reward answers that correspond to work at the analysis, synthesis, or evaluation levels. It is interesting to note that when given the opportunity to choose between a forty-item multiple-choice test and a problem-essay test, about 90 per cent of the students selected the multiple-choice test, although retesting revealed that the level of performance for the same students was higher on the problem-essay test. The students generally avoided the mathematical questions on the problem-essay tests (Winter, 1967). The Project Physics course exhibits diversity even in its tests.

Although some Project Physics tests could be used by some teachers as unit tests, in general, their intrinsic relationship to the Project Physics course and the emphasis upon the history of science would make them unsuitable for general use with other textbooks. As with the PSSC tests, it seems that the teacher of any high school physics course could profitably study these tests and include some items of similar construction in his own tests.

REFERENCES
Bloom, B. S., ed., *Taxonomy of Educational Objectives, Handbook I: The Cognitive Domain* (New York: David McKay Co., Inc., 1956).
Nedelsky, Leo, *Science Teaching and Testing* (New York: Harcourt Brace Jovanovich, Inc., 1965), p. 142.
Winter, Stephen S., and Wayne W. Welch, Achievement Testing Program of Project Physics, *The Physics Teacher*, 5:229-231 (May 1967).

Cooperative Science Tests Physics, Cooperative Test Division, Educational Testing Service, Princeton, N.J., 1963.

These tests are part of a series that includes tests ranging from grade 7-9 general science, up through biology, chemistry, and physics. The physics test is intended for use in grades 10-12, which is consistent with the placement of physics in the high school curriculum. Development of this series of tests was begun in 1958, and the final versions were produced in 1962.

Two forms, A and B, are available for the physics test. Each test consists of two parts. Part I has sixty items that are intended to test knowledge of the content, concepts, and principles of physics. Part II has fifty-five items dealing with laboratory equipment and problems related to laboratory situations. Each item consists of a stem with five possible responses, and the student is to select the correct response from these. Separate answer sheets must be used with these tests. A working time of eighty minutes is required for the test, forty minutes for each part.

The following discussion is based upon form B of the test. Although these tests were designed to meet the needs of updated science programs,[1] the physics test deals with what may be described as traditional physics, and the PSSC course seems to have had little influence upon the test. There are four items involving fluid pressure or buoyancy, eight items involving image formation with thin lenses, and one item dealing with simple machines. As far as content areas are concerned, electricity seems to be disproportionately represented; there are about nineteen electricity-related items on the entire test. This weighting seems too great in terms of the few items that are used for concepts such as gravitational potential energy, momentum, or motion. Only item 29 of Part I was found to be inaccurate; answers B and C are both correct for this item. Item 8 of Part II seems somewhat ambiguous, but this is probably a result of attempting to construct an evaluation item in a multiple-choice format. Items 7–9 of Part I deal with observations of a water tank in a moving vehicle and are excellent items; however, they could be considered dated since the vehicle described is a parlor car on a passenger train.

The test handbook provides information on the tests' reliability and standard error of measurement, based upon a sample of 450 examinees. Normative information on both national and urban samples are included. A discussion of how the test scores may be used is included. An interesting feature in the handbook is a breakdown of the test items into four skill levels: knowledge, comprehension, application, and analysis and evaluation. Some of the items' classifications could be very different depending upon the teacher's class and laboratory topics, but this is still a desirable feature on a test.

The physics teacher would be most likely to use this test as part of a school-wide, long-term testing program. As such, it could be useful, especially if local norms are established as described in the handbook. The teacher who is teaching a traditional physics course, as opposed to Project Physics or PSSC, might wish to use Part I for his own information. However, the inclusion in a teacher-made test of items involving actual experiments and the manipulation of laboratory apparatus could result in a test of equal or greater effectiveness as compared to a paper-and-pencil instrument such as Part II of this test.

---

[1] Handbook: Cooperative Science Tests. Educational Testing Service, Princeton, N.J., 1964, page 57.

Other physics tests that are currently available:
  Tests of the Physical Science Study Committee
    Published by the Cooperative Test Division, Educational Testing Service.
  Dunning-Abeles Physics Test, Harcourt Brace Javonovich, Inc., 1967

Standardized Chemistry tests that are currently available:
  ACS-NSTA Cooperative High School Tests
    A. Regular High School Chemistry Tests
    B. Advanced High School Chemistry Tests
    Produced by the Examinations Committee-ACS
  The Anderson Chemistry Test, Harcourt Brace Jovanovich, Inc.
  The Cooperative Chemistry Test, Educational Testing Service, Princeton, New Jersey

ADDRESSES

Educational Testing Service
20 Nassau Street
Princeton, New Jersey

Examination Committee-ACS
University of South Florida
Tampa, Florida 33620

Utilizing Instructional Media

How will today's teaching differ from yesterday's? Will the students be motivated to seek answers for questions beyond those posed in the classroom? Could the lecture portion be read for greater student profit? How realistic is the teaching in terms of its future application? How can the teacher alter the pattern of lecture that has been ingrained in him through his years of schooling? What can he do to instill his personal excitement for the subject? Can the efficiency of learning by the student be increased and if so, how? The answers to these questions are not simple ones and will require a lifetime of trial and modification by the teacher. The answers must be sought so that the students with whom the teacher comes in contact may learn at the maximum rate.

Instructional technology is alive and will be working in more schools to a greater degree in the future. The hardware, consisting of projectors, recorders, television, and the like, is available to make learning an exciting experience as opposed to the rather dull exposure that most students have to science, chemistry, and physics in particular. The software of transparencies, slides, tapes, and the like, is available from many publishers to diversify instruction. Instructional technology can assist the teacher in creating a learning environment in which the students will learn more material in a meaningful fashion than was previously thought possible. The key to implementation is the teacher's willingness to exert the necessary energy for assimilating the myriad possibilities and incorporating them within a varied instructional format.

To utilize the available media in the classroom the teacher must be aware of what exists with respect to hardware and software. He must also have a detailed conceptual breakdown of the subject to be taught. This will include a list of concepts, their order for efficient learning, and a statement of objectives for the course. It is impossible to chart an instructional sequence if the teacher has no specified goals. It is also impossible for the teacher to use those materials prepared by others with the same degree of effectiveness as he would his own. The production of materials becomes a part of oneself, which makes the teacher more contagiously enthusiastic.

The limited scope of this book cannot provide teachers with all the directions for using the obtainable hardware or for preparing the software. An attempt will be made to acquaint the teacher with the audiovisual resources that are available and to list specific references that will assist the teacher in the use of each type of audiovisual material.

The teacher must examine the instructional format of the classroom against the environmental conditioning that the student receives. The lecture as an effective method of conveying information has long been dead. New ideas must be implemented in classroom instruction. The students are constantly bombarded by radio, television, and other sources with words and ideas to the point that they may no longer hear what is being said. The teacher will have to be an actor of the highest caliber to sustain the students' attention while using lecture to describe the intricacies of the electron and other subatomic particles that the students have failed to conceptualize.

For the students to meaningfully incorporate what they are hearing, any presentation must be supplemented with visual materials that will assist them in relating the new concept to that which is familiar to them. The chemistry or physics teacher may be tempted to ignore this because of the higher quality student who typically enrolls in his course. The teacher should recall that one of the reasons that the student has the higher grades may be his ability to memorize vast quantities of useful information that was quickly forgotten. Hopefully, the science teacher will want his students to do more than memorize. He should instead strive to instill in his students the desire to learn by giving them a frame of reference for the incorporation of science concepts and the skills necessary for such incorporation.

Ideally the function of the teacher's instruction will be to provide a multisensory input to the student's cognitive structure. If the teacher is striving to have each student learn to a certain level of understanding, it will then be necessary to provide the student with the information or input in a usable form. Some students relate to actual laboratory or field experiences. Others need only to be shown models along with the appropriate explanation, and a textbook presentation may be sufficient for yet other students. Some students will even require a combination of all possible inputs. Motivation to learn and increased learning may take place by requiring a student to look at a particular concept from another input source. If a student is accustomed to verbal descriptions and he is then required to formulate his own descriptions from the observations of a model of a demonstration, he may take increased interest in the subject. Possibly the new mode of presentation may allow an increased efficiency of learning by the student.

The different modes for the presentation of concepts that have been formulated by the teacher will actually allow for individualized instruction of the highest order. Each student will be able to learn at his own rate and will be allowed to repeat units of instruction that he may have missed initially. The student would also be allowed to select the particular method of instruction that most appeals to him, thus increasing both the efficiency and interest. Those skills that must be learned in each subject in a rote fashion can be presented in a programmed format that will give the student the practice he needs in the shortest time possible. The programmed format also provides additional class time for the more diverse concepts or presentations.

The student must have concrete and physical references to which he can relate new material or he will fail to learn it in a meaningful and usable manner. Verbal descriptions cannot take the place of physical realities in the minds of most students. The teacher must strive to keep a reasonable physical representation of the concept before the student at all times. He should also have the student actively explore that reality to provide for a higher level of understanding. Which learning is better: to be told about the properties of a pendulum or to discover the different properties of the pendulum by self-experimentation?

The instructional media that are available to chemistry or physics teachers may be classified as follows: (1) actual student laboratory experience in handling matter; (2) 3-D models; (3) 2-D models in the form of pictures, films, and drawings; and (4) mechanical presentation, e.g., tapes, programs, and the like. The actual selection of media will depend upon the instructor, his students, and the concept that is to be learned. Each medium has its limitations and strengths; the teacher must be aware of these to provide optimum instruction.

The contributions that media can make to instruction are many and varied. The instructor must select the media with the objectives of instruction as his primary criteria. The use of two or more media will multiply the effectiveness of each and will reduce the limitations of the individual medium by a corresponding amount. Media can function in the following manner for the instructor:

1. To provide a visual model. If the student can see what the instructor is talking about, what is said will be remembered longer than if the student only heard what was said.
2. To provide information that is inaccessible in any other form. Some experiments are too hazardous for the student or even the instructor to attempt in the laboratory. Films or slides will be used in these cases to bring into the classroom the information that is unavailable in other formats.
3. To provide broader coverage. Television is being used to show the details of phenomena that are confined to small areas to large groups of students. It is also being used to bring phenomena across the country at the time it is happening. The quantities of

materials that the instructor must use for the demonstration are reduced to safe levels, and the student is still able to see the materials in as great or greater detail than he would otherwise have.
4. To provide for the learning of basic skills. Programmed instruction is used as a tool to raise the proficiency of some individuals to an acceptable level. Some students lack the necessary mathematical skills to analyze data or solve the problems that are encountered in chemistry or physics.
5. To attract attention. Presentations of equipment or pictures in the classroom or in display cases can arouse the curiosity of the students for further investigation. These displays may also be used to motivate the students who are not enrolled in the course to do so in the future. Current activities in chemistry and physics may be featured to provide a general education to those viewing the display.
6. To provide before and after comparisons. Film loops or snapshots may be used by the students to detect subtle changes that may have occurred since an experiment was begun.
7. To slow down, speed up, or stop the action on a particular sequence of activity. Films and pictures can provide the student with the input necessary to make observations about the action that might not have been otherwise possible.
8. To show those parts, mechanisms, or processes that cannot normally be viewed. By using models, pictures, and films, the workings of mechanical devices, such as engines, can be shown in great detail. Cutaway diagrams allow the student to see the relationship among the various parts of a machine that could not have been seen any other way. Animation can provide a means of using models to demonstrate the motion and interactions on a macroscopic scale that occur on a submicroscopic scale.
9. To update materials. Transparencies or handouts of recent scientific articles enable the student to learn about current changes.

The limitations of the media depend upon the student perception of the media. If the student is working with matter in a 1:1 ratio there should be little chance for a misconception of the physical properties. If the student is studying the effects of gravity on falling objects by actually dropping objects and measuring distances and times, his sensory input of the situation is real. If the student studies the concept from films or pictures, part of the input is reduced although the conclusions may be the same. One dimension has been removed from the student and the student will thus often receive a slightly distorted impression of what is actually happening.

A model of an atom in three dimensions gives the student a physical representation of the various shapes and sizes of the subatomic particles. The teacher must be aware that misconceptions often occur such as the distance between the nucleus and the electron as well as the size of the electron with respect to that of the nucleus. The model is three-dimensional as would be an actual view of an atom. However, using macroscopic materials to represent a microscopic reality will cause distortions. The teacher will often fail to inform the students of the errors that can occur by relying solely on this model because he has come to accept the misconceptions without thinking about them. The student viewing the model for the first time will accept it as real unless he is informed of the distortions.

Pictures, either stills or movies, will also distort the input to the student. They will focus the students' attention better than verbal inputs. The questions to be answered by the pictures are not always obvious because the student does not have the full opportunity to view the matter as it should be viewed. Supplementary information must be provided to avoid the formation of misconceptions by the students.

Working miniatures may also create wrong impressions because some forces may not be scaled up or down and still retain the same physical relationships to one another.

Presenting a visual item in color rather than in black and white will also influence the way in which the student will perceive the material. There are subtleties in observing changes in chemical reactions that can only be appreciated if the reaction is being observed in color. The same effect would be lost if viewed in black and white.

A visual presentation is obviously to be preferred to a verbal presentation, particularly if the student has never seen the phenomenon or object being discussed. In the teaching of chemistry and physics where relationships among different forms of matter and energy are so important, the teacher cannot afford to rely entirely on verbal descriptions.

Software

The teacher has a choice of a variety of sensory inputs to the student. Some of these are already available in the school classroom or nearby, although other materials must be secured from commercial sources. The teacher must usually select from among the following resources as he prepares for each class.

1. Laboratory
2. Chalkboard
3. Demonstrations
4. Models
5. Field trips

6. Transparencies
7. Bulletin board
8. Photographs and slides
9. Film loops
10. Films
11. Filmstrips
12. Television
13. Audio tapes
14. Computer-assisted instruction

LABORATORY

The effective use of the laboratory as part of the instructional process is discussed in Part 4. It is expected that a significant percentage of the classroom time would be spent in the laboratory investigating ideas using the masses and energies being discussed. Only when these materials may be dangerous for the student to handle should the use of other media be encouraged. The quantity of time spent in directly handling the materials might be excessive, and a brief exposure via some other media would be desirable.

CHALKBOARD

The chalkboard is the primary visual aid for most teachers. When all else fails, the teacher can use his skills in drawing to illustrate mechanisms or convey materials not otherwise available from a common source. The wise use of colored chalk will assist the teacher in illustrating ideas and setting off different functional areas within an overall drawing. The chalkboard may be used when time does not permit the teacher to construct other media. Problem-solving activities require that a medium be used that will allow all of the students to record the results as well as to see the step-by-step solution. Problems presented by the student can seldom be anticipated in advance, and so this medium will continue to be used extensively. The chalkboard has its drawbacks because teachers tend to rely on it for the bulk of their presentations. Students are required to copy procedures introducing a possible source of error. Since instructors usually continue to talk while the student is copying material from the board, important explanations are frequently missed. If the instructor prepares a handout and a transparency, the flow of instruction will increase. Teachers are also not the artists they claim to be. Many concepts are missed or misconstrued because of inaccurate detail in drawings. Rather than relying on the board to demonstrate an idea, the teacher should use models and demonstrations to give more detail than is possible on the board.

DEMONSTRATIONS

Demonstrations serve to introduce a problem area with a minimum amount of confusion. The students can see what is happening and if necessary can record data to be used in the development of answers to problems. The input is three dimensional instead of two dimensional so the student has a concrete referent in the real world and can incorporate related material in a more meaningful fashion. The demonstrations must be carefully chosen and used when the instructor is sure they will work as he wants. Demonstrations have been included with the common concepts earlier in this book (Parts 2 and 3), and the teacher is encouraged to use them if possible. A disadvantage to the use of demonstrations is that all students will not be able to see equally well depending upon the number in the class and their relative positions to the teacher. Demonstrations are also time-consuming, and the teacher should be able to justify their use on the basis of expected benefits to the students.

MODELS

Models can function as permanent displays in the classroom with an appropriate guide sheet. They may also be used in conjunction with class discussion when other media cannot produce the same effect. For example, the structure of the atom or crystals can be presented more effectively with models than with two-dimensional media. Scale models may also be included in this category as it is impractical to have large items such as generators or airplane wings in the classroom. The students still obtain a three-dimensional feel for the object and its functional parts, which would be impossible from two-dimensional media. The objection to models is that they may need a large storage area as the teacher continues to acquire them.

FIELD TRIPS

Field trips may be used to emphasize industrial processes and the scale at which these operations are carried out. Nearby museums containing displays that can be related to the topics being studied generate enthusiasm and stimulate class discussion. The field trip is time consuming and may require that all teachers cooperate in allowing the students class time to participate. The instructor in charge is faced with increased responsibility for the students and for the educational objectives to be achieved by the field trips.

TRANSPARENCIES

Transparencies are used by many teachers as a means of reducing the amount of time that is required in the presentation of material. These may be prepared in advance or secured from a commercial source. The teacher can increase the effectiveness even more if the student is provided with copies of the same information. Color may be added to cutaway drawings and, by combining several transparencies, different areas of the same basic mechanism may be presented and emphasized. Transparencies are usually inexpensive and easy to prepare even to the point of preparing them as the class discussion proceeds. They are used in this fashion when the instructor wishes to face the student while discussing the solution to the problem. The difficulties of transparency use are limited to using a two-dimensional mode to depict a three-dimensional reality. For problem solving or general written exchange, it can be as effective or more effective than any other medium.

BULLETIN BOARDS

Bulletin boards should be used by the teacher to present related material or as a means of arousing student interest. Items of current events, which may be clipped from newspaper or magazines can keep students abreast of what is happening in science. This requires that the teacher read widely in popular and scientific journals. Making articles available in the classroom will enhance the probability that the student will spend some time reading them. The techniques for layout and the principles of good audiovisual communications via the bulletin board are described in references listed at the end of this chapter. The teacher is encouraged to secure some of these books and to apply their contents to his individual situation. Bulletin boards should be changed on a regular schedule to prevent the material from becoming outdated. A disadvantage of this medium is that the amount of material that may be presented is limited by the space available. This may be compensated for by assigning the class or a small group of students to develop a display that will report the findings of their investigations so that others may have an opportunity to examine them. The science teacher should not limit this form of presentation of material to the bulletin board. Any bare wall space, as long as it does not conflict with school policy, may be used to post articles of interest to large numbers of students. The door to the classroom and the adjacent walls may also be used if masking tape or another type of nondestructive material is used to secure material to the wall. This will enable the teacher to inform students who are not enrolled in science courses of developments that will affect their lives.

PHOTOGRAPHS AND SLIDES

Photographs should be used when it is impossible to view the real object. Their cost will limit the use of photographs to black and white so the teacher must consider what will be lost in the process. Photographs should be used in conjunction with other materials such as tapes or oral presentations to develop a frame of reference for the student. They make excellent material for bulletin board displays when they are accompanied by the appropriate narrative. Photographs also function to record experimental set-ups for inclusion in group reports. The teacher should have access to a Polaroid camera that may be used in the classroom to record data that the students might want to reexamine at a later time.

Slide presentations are similar to photographs in their use, although the teacher does have the advantage of color. Slide presentations will be particularly effective if used when long narratives are planned to discuss a particular industrial process or to explain the functioning of larger pieces of equipment. They may be copies of models that must necessarily be small but that may be enlarged for ease of student viewing. Slides are relatively inexpensive and may be purchased from commercial suppliers or made by the teacher. Again the techniques required to produce effective slides are beyond the scope of this book, and the teacher is encouraged to examine some of the literature cited at the end of this chapter.

FILM LOOPS

Film loops in an 8mm film loop are available to the teacher for a variety of topics from safety to experiments. Film loops come in either color or black and white and last from two minutes to eight minutes. They are usually limited to a single concept that may be repeated several times during a single class session. The viewing must usually be restricted to ten or fewer students as 8mm film loops do not show well for large groups. They are effective in presenting concrete examples of physical properties or of chemical reactions that may be too hazardous for the students to examine directly. Student time is saved by the use of these loops because the student need not set up equipment and develop skills for handling the equipment. The student may also examine each film loop several times rather than once for a laboratory procedure. By using stop action, the student may collect data directly from the film loop to help confirm a hypothesis that was proposed earlier. Teachers should examine some of the film loops that are currently available and evaluate the role they could play in helping the students achieve the stated objectives. Most companies will provide a ten to twenty day examination period before purchase. Lists of suppliers may be found in the Appendix. The ambitious teacher may even try producing his own film loops, although the time necessary to produce a satisfactory film is extensive. If the school system has a large audiovisual support system for the teachers, the system might be able to assist the teacher in the production of the film loop. The personnel could provide the techniques and the expertise in the audiovisual realm that the teacher probably lacks.

FILMS

A large number of 16mm films are available at no cost from manufacturers or for rent from universities or other commercial suppliers. Films have formed the backbone of most audiovisual systems in the public schools for many years. They present an opportunity for the students to view people, places, and events that would ordinarily not be available. It should be remembered that the films become outdated and that not all films will assist the teacher in achieving the objectives for his students. Frequently the films ordered will not be available when they would be most effective and the teacher cannot always anticipate when each class will be ready to use them. For the first few years the teacher is usually experimenting with the curriculum for each course, so that when films must be ordered six months in advance the date chosen may not coincide with the date that would be most effective. Films that are used should be rated and a record of responses and uses should be made so that further use of the film will be given careful consideration. Other science teachers in the school could be contacted to determine if they have used the film and what their impressions were. Ordering films in the larger school systems will be handled through the librarian or the audiovisual department. The teacher will probably be able to secure catalogues from these departments and submit requests to the people in charge. Those in charge will then order the film and notify the teacher when it arrives. If the teacher must do the ordering himself, he should start with a set of up-to-date catalogues. These should be obtained from universities in the state where the teacher is working and from some of the larger commercial suppliers. The addresses of both may be found in the Appendix. Letters should be sent to these sources containing the title, catalogue number, and play dates for the films desired. Alternate play dates should also be included if the teacher believes that the film would be a valuable aid irrespective of when it will be shown.

FILMSTRIPS

Filmstrips, like slides, may be used to illustrate those phenomena that would not be distorted by the stop action. They are cheaper than films and may be purchased for repeated use if they prove effective in the curriculum. Filmstrips are usually bought from commercial sources, since few teachers prepare film strips themselves. The teacher must evaluate any film strips purchased by the school on a continuous basis so the content presented will not be outdated by later information. When film strips are combined with audio tapes, they function very well to individualize instruction in a remedial capacity or as an independent study area.

TELEVISION

The use of television in the science classroom will be increasing in the next few years. The cost of color units is now within the reach of most school systems, which will give the television tape the same advantages of film loops, slides, or laboratory work. The teacher will be able to prepare presentations of physical phenomena without worrying about the excessive costs involved with the other formats. Video tape may be reused many times without any change in the quality of the picture. Views of microscopic events may be enlarged so that the entire class can simultaneously observe the processes in action. The tape can also eliminate the unexpected happenings that frequently occur with demonstrations or laboratory work. The portability of the latest videotape equipment gives the teacher the flexibility of bringing community resources into the classroom thus eliminating, to some extent, the need to take field trips. Visiting scientists or other resource people who would not be available for a full day of contact with the students may be utilized. The teacher can also utilize television as a self-evaluation instrument. Various instructional techniques may be examined for effectiveness and for skill development. Video tapes of particular classes in progress allow the teacher to evaluate the students' work from a different perspective. The actions of all the students may be scrutinized instead of the limited number that the teacher usually sees. The hazards of using television are centered in the fact that students have been exposed to a wide variety and quality of television programs for several years. They can recognize poor quality when they see it and may respond accordingly. The teacher will have to be extremely critical of the material that he presents in this format. Another disadvantage of using television is the three-dimensional projection into two dimensions of television. The students are familiar with television but that does not mean that a physical reality is completely understood if this is the only format in which it is viewed.

AUDIO TAPES

Audio tapes complete the line of software that will be available to most teachers. These come in two formats; the cartridge tape and the reel-to-reel. For classroom use by individual students, the cartridge tape will be the accepted format because of the simplicity of its use. It is practically impossible for the students to ruin one of these tapes. The machines for playing cartridge tapes may be purchased with a play-only feature that would prevent the students from erasing the tapes. If top quality is desired, the reel-to-reel tape may be the more appropriate format. Many schools will have this type of machine and may have good tapes in the library. For most use, however, reel-to-reel tapes should be recorded in the cartridge form for student use. Group discussions may be recorded for transcribing when problem solving is being conducted by the group. Directions for the use of this equipment may be presented by the use of these tapes. Audio inputs to the student may be used when the visual input from texts and pictures is insufficient to convey the concept to the student. The audio tape is one of the formats that is helping to individualize instruction by allowing the student to chose the format of instruction that he desires and providing that the chosen mode of instruction will insure the success that is desired by both the student and the teacher.

COMPUTER-ASSISTED INSTRUCTION

Teachers may also have some form of computer-assisted instruction available for their use. To date, computer-assisted instruction has been shown to be effective in drill work and in some other areas of science instruction. Its development will be retarded by the cost of the equipment used and the expense involved in writing the programs to instruct the students with the flexibility the teacher presently possesses. The teacher should use computer-assisted instruction, if available, to provide the student with this experience but should not use it on a continuing basis until further testing of the educational results have been completed.

HARDWARE

The teacher will have a large variety of equipment at his disposal that may be used in the individual classroom. Much of the larger equipment will only be available from the audiovisual department when it is requested in advance. This equipment will come complete with operator, and the only responsibility the teacher will have is to order the software. Other school systems will not have this service, and the teacher will have to do the work himself. Some of the hardware will be located permanently in the classroom, and the teacher will be responsible for the selection of the equipment that will do the best job. The teacher should not order any equipment without discussing the purchase with the audiovisual director if one is available; he will be more knowledgeable about the current quality of the equipment being considered. From the discussion the audiovisual director will also be able to determine the objectives that are to be served by the equipment and his recommendation should carry maximum weight.

Teachers in other school systems may have to chose equipment and operate it themselves. This section is designed to provide criteria for the selection of the equipment as well as a few of the basic operating instructions. It is impossible in the short space of this chapter to present a comprehensive survey of the equipment. The teacher is referred to the bibliography at the end of the chapter in selecting books that cover the equipment and procedures for its use in greater detail.

The teacher will find that in assimilating the equipment that is necessary to the implementation of a multimedia science curriculum he must choose the following pieces of equipment:

1. 16mm projector
2. 8mm projector
3. Filmstrip projector
4. Slide projector
5. Overhead projector
6. Opaque projector
7. Cassette tape recorders
8. Reel-to-reel tape recorders
9. Television equipment
10. Cameras
11. Screens
12. Calculators

16mm PROJECTOR

The teacher will be confronted with a choice of projectors; most of them will perform the job adequately. The primary feature of concern is that of the lamp used and the cooling system for that lamp. A 500-watt bulb is the minimum acceptable to insure adequate light on the screen. The cooling system is harder to rate, but for the larger bulbs there should be a greater amount of air moving around the bulb, which will be indicated to some extent by a larger opening in the top of the projector through which the heat may be conducted. The magazine *Audiovisual Instruction*, which regularly features product evaluations, may be consulted.

There is also a choice between manual threading and self-threading projectors. The manual machines may be a little harder to thread but they offer a degree of flexibility with respect to being able to start the film anywhere along its length. The self-threading machines can also be used this way but it is more difficult. If a trained operator is to use the machine the manual threading will provide a savings on the initial outlay. If students are expected to operate these machines (and they usually are not), a self-threading machine would be desirable.

A trained operator will bring the machine to the room, set it up, show the film, rewind the film, return the machine to its proper location, and see that the film is returned to the company from which it was rented. Otherwise the teacher must do these tasks by himself. In so doing, he must attend to the following details:

1. Procuring the equipment
2. Setting it up
3. Providing for total viewing
4. Rewinding the film
5. Returning the film

The equipment should be reserved in advance as the teacher probably will not have a projector assigned to his room on a permanent basis. This requires that the other teachers know of the intended use so that conflicts may be prevented. The projector should come complete with a projection stand to minimize the problem of projecting the picture from an adequate base. The stand should also have an extension cord so that one will not have to be found when it is determined that the projector cord will not reach the nearest power outlet.

The projector should be located in the room on a line perpendicular to the face of the screen. The screen will usually be located in either the center of the front of the room or to one side of the corner. The ideal location for the projector is at the center of the room, as this will provide for the most comfortable viewing by the students. It will also allow all of the students to see the picture without any distortion effects that result from being too close and to one side of the screen.

To thread the film, the teacher should first look for a set of directions for each machine. Instructions are usually located on the machine because of their frequent use. It may be necessary to look inside the machine cover or possibly under the machine for the instructions. A diagram of the film threading will be shown and the teacher must be certain before showing a film that the film is threaded according to the diagram. Other problems that are often encountered include: having the rewind mechanism engaged, having too much slack in the film, not having the film locked in position with the appropriate tension behind the lens, and improperly attaching the film to the take-up reel.

The teacher should become proficient in the operation of the machine before setting it up in the classroom. If necessary, he should consult with an experienced teacher and try threading and running the machine under his supervision. Everything should be ready to run when the students arrive in the classroom. This includes threading, framing, and focusing the picture so that the full screen is used.

The teacher should provide for total viewing by centering the projector in the room to use the largest area of the screen without running the picture off the edges. The quantity of light in the room can be regulated by the shades in the room. If there are no shades, arrangements should be made to use a room that will give the correct lighting. When the students are to take notes, a low light level should be maintained.

Rewinding a film proceeds when the film has run all the way through the machine or the teacher has stopped the machine at a preselected point in the film. In the latter case, the film must be removed from the threading mechanism, drawn up tight on the reel, the rewind mechanism engaged, and the motor started. In the former case the film end should be run from the take-up reel to the other reel in a direct line. It should be attached to the empty reel in the same manner in which it was taken off so it will be ready for storage. A rewind switch must be engaged and the film will then rewind by turning the machine on without the use of the lamp. Many machines are equipped with a reverse mechanism, but this should not be used for rewinding film as it will produce more wear on the film and increase the probability of breakage.

Where and when to rewind the film must fit in with the instructional activities of the day. It is best to bring out the strong points of the film immediately with the students while the experience is still fresh in their minds. On the other hand, if the next class will be viewing the same film, it should be rewound before the end of the hour and made ready for showing. The teacher must make these decisions based on the contents of the film and the length of time required to show it.

Many companies will request that a film not be rewound before its return. They will usually want to check to make sure that the film has not been damaged while it is in the teacher's possession. The teacher will occasionally have to decide to ignore this if the school does not have a reel that exactly matches the one on which the film came. Some suppliers will request that their films be insured for a specified amount before return; the teacher is encouraged to honor these requests. Labels and containers are provided for the film, so all the teacher has to do is put the film back in the can, fill out the return label, attach the label to the box, and attach sufficient postage. The secretaries in the main school office will usually take care of the last item.

8mm PROJECTORS

Presently there are three manufacturers of 8mm projectors for film loops: Technicolor, Kodak, and A. B. Dick Company. For teachers preparing their own films and using reel-to-reel films, the considerations that apply to the 16mm projectors apply here. The 8mm projector has been used increasingly because it will show films to small numbers of students in an individualized instruction format. These machines must therefore be relatively foolproof. The cartridges that have resulted fit this criteria well. The most abundant if not most popular projector is the Technicolor projector that offers a variety of options from simple projection of the film loop to projection with sound. There are already more film loops in the Technicolor format than either of the other two formats. However, most film loop producers make them available in the first two formats. At present, the only company producing the last format is the A. B. Dick Company, which manufactures a number of films for use in chemistry and physics. The teacher should, with the aid of his colleagues, select one format and use it exclusively since the cartridges are not interchangeable. This decision will have already been made at most schools.

The 8mm film is used to present a single concept or to demonstrate a single procedure. As such, it is usually not used for large group instruction because the lamps in these machines do not have sufficient power to project onto large areas and maintain the desired clarity. One or more of these machines should be located in the room in such a way that the students may use them individually or in small groups. Carrels are provided for the use of the films, and a small screen allows the student to see it clearly without the room having to be completely dark. The teacher will have to check the machines periodically to determine their operability. The film loops will also have to be examined because the students will often leave the film loop at some point other than the beginning. The film loops should be run to the end to prepare for the next student. Instructions should be given to the student using the films indicating that each individual is expected to return the film loop to its beginning. Instructions for the use of the machine should also be posted next to the machine. The following is an example of the list that might be included with a Technicolor projector.

1. Insert the cartridge in the slot at the rear of the machine.
2. Turn the machine on with the switch on top of the machine.
3. Focus by twisting the lens cover on the front of the machine.
4. Frame the picture by rotating the knob on top of the machine.
5. View the film completely as many times as is necessary.
6. Turn the machine off when the film is at the end.
7. Remove the cartridge and return it to the designated location.

FILMSTRIP PROJECTOR

The filmstrip projector should be selected with respect to ease of operation and lamp brightness. The intended use of class or individualized instruction should be a factor in the selection. The number of filmstrips for chemistry and physics that are now being produced and shown is not as great as in some other areas. The filmstrip projector will probably only be used on a limited number of occasions and will not be a permanent part of the equipment assigned to the room.

Threading the individual machines is not standardized, and the instructor should either read the directions or seek help from someone who knows how to operate the machine. Student use of these machines will also be limited since most still color presentations will utilize slides rather than filmstrips.

SLIDE PROJECTORS

This machine should be selected with respect to ease of operation, size of lamp, and versatility. The machine should be easy for both the instructor and the student to use. Many of the slide projectors on the market require the slides to be mounted in cartridges. When the cartridges are in place, flicking on a switch will advance the slides one at a time. As long as the slides are not resequenced, the presentation remains the same. The teacher should try as many of the different types of cartridges and machines as possible. Most cartridges are either circular or longitudinal with the difference accounting for the number of slides that may be packed together; circular cartridges that will hold about 120 slides are preferred. Most presentations do require this number of slides so each teacher will have to decide. The circular cartridge may be segmented and spaced forward to the appropriate place. The teacher must also consider the machine with respect to student usage. It is desirable to have a machine that requires the minimum amount of student thought in its operation. At the maximum, the student should be only required to turn it on, focus it, and change from one slide to another. The lamp should be at least 500 watts for individual viewing and 1,000 watts for group viewing where projection over long distances is required. Versatility in some machines allows for the use of the slides in cartridges, individual slide viewing, and an attachment for the use of filmstrips. The filmstrip attachment eliminates the need for a filmstrip projector as such. Other features include sychronization with a tape recorder so that a complete audio and visual presentation may be presented to the student at any time.

Using this machine in instructing a large group requires the same considerations as those described for using a 16mm projector. If the teacher does not have a machine assigned to him on a permanent basis, the appropriate arrangements should be made so the projector will be available on the needed date. Slide sequencing must be done in advance to utilize all of the class time for discussion rather than preparation.

These machines are usually more complicated than the film loop projectors. The teacher should train the students in the use of the equipment to minimize the amount of accidental damage that may be done. Once the students have been trained, a set of directions for the use of the machines posted in the carrels will serve to remind them of the steps to be taken. The slides may be coupled with written material or an audio presentation may accompany them for diversification.

OVERHEAD PROJECTOR

This piece of equipment will allow the teacher to illustrate spontaneously where other equipment will not. A large number of projectors are available with a number of accessories. The first consideration should again be for lamp size because the more power available the better is the viewing from the rear of the room. The 800-1,000 watt lamps are considered minimum although smaller lamps are available. To prevent damage to the transparencies used, an adequate cooling system must be utilized that will vary from machine to machine. The lamp itself will be located either in the base below the platform or in the head of the projector. Both locations serve the same purpose, but the larger wattage bulb may be used only if the bulb is in the base of the machine. Some machines come with acetate roll attachments, which are good for a continuous presentation or when it is desired to return to a previously covered problem. This feature is good to have if money is not a major consideration.

This projector is used primarily to present material for the students' consideration. It allows the teacher to view the class while writing or talking, which tends to prevent misunderstanding. When transparencies are prepared in advance, copies should be mimeographed and handed out to the students to allow them to listen to what the teacher has to say. Students may use this machine in class discussion when individual problems are assigned to groups of students who are asked to present their solutions to the entire class. It is relatively easy to give each group a grease pencil and a sheet of acetate on which to record the solutions to their problems.

OPAQUE PROJECTOR

This projector will have only a limited use and therefore the teacher will probably be using one shared by several teachers. Desirable features of this apparatus include ease of inserting the material to be shown and the amount of light that is actually projected on the screen. A 1,000 watt bulb is a minimum figure and a high wattage should be used if possible.

The students will seldom use this machine, and the teacher will use it only to present material that has not been copied into another format. Pictures may be shown from magazines and books during a class presentation. Materials to be shown should be inserted in the bottom of the machine and positioned to show all of the material on the screen. The machine should be mounted on a projection stand and located with respect to the screen to provide for maximum viewing. The room will frequently have to be darkened to provide the clearest picture on the screen. Focusing is a simple matter of rotating the appropriate knob near the lens. Some machines have a choice of light outputs, and the teacher should select the one that provides the sharpest, clearest picture.

The teacher should be sure to reserve this machine in advance and should then promptly return it when the showing is complete. Experience will indicate the value that this machine will have to the individual teacher.

CASSETTE TAPE RECORDERS

The choice in these machines is varied and the objectives should dictate which one should be selected. The first choice is between a play only and a play and record machine. The play only machine is ideal for student use when no recording by the students is required. The students will not be able to erase tapes, which assures the teacher that the materials provided on the tapes will remain intact. These machines can be battery or a.c. operated. The battery models are more portable, but the cost of the battery can offset these benefits particularly if the machine is to be used where electric power is available. The battery operated machine is usually the least expensive.

Features on any recorder should include forward, fast forward, fast reverse, and stop modes. The combination machine gives a greater diversity in places where it may be used by the student. The quality of sound reproduction on these machines is usually adequate but the teacher should check this out before ordering any. Voice reproduction is better than music reproduction on the smaller recorders; this should be kept in mind when preparing scripts. The electric model is restricted to use where power is available but this will be no problem in most school situations.

The play and record machine may be used by the student or students to listen to directions or record group discussions. The play portion should be of sufficient quality to provide adequate voice reproduction for easy listening with either the speaker or the headphones. The record portion should be easily operated by the students and the teacher and should give satisfactory voice reproduction. Features should include those mentioned earlier. An additional desirable feature is a footage meter to permit a student to start at a predetermined point in the tape without much starting and stopping that produces wear on the tape. This feature will cost extra and if each tape is continuous from start to finish it may not be necessary to equip each machine with this feature.

The master recording machine on which the teacher will record programs for use by the students on other machines should include all of the features mentioned previously. The footage meter will be particularly helpful when the teacher must follow a script or allow time for student responses. This machine should also incorporate some type of tonal quality adjustment to filter out background and machine noise. The microphone should be able to pick up conversations at a distance of fifteen feet or more and at the same time give excellent results when the speaker is close to the microphone. A variety of tape speeds is not essential as most of the machines that the students use will be playing the tape at 1-7/8 feet per minute. A jack should be available so that the tape may be played from this machine and recorded on another machine without allowing for added noise by playing it out in the room. Stereo is not a requirement unless this machine is to be used by other departments in the school. Usually a stereo cassette recorder will give adequate music as well as voice reproduction. The tapes must be played on a higher quality player to be fully utilized. A good tape on a cheaper machine will not sound as well as it could.

Directions for the use of these machines should be taped to the lid where the cassette tape is inserted. The directions listed for play only machines will consist of inserting and removing tapes, starting and stopping the tape, and rewinding the tape to its original position. Directions for the record and play machines should separate the play from the record modes as the record part will often not be used. In the play mode the directions are the same as before with the added caution not to engage the record mechanism as the tape would be erased if this were to occur. To record, the student should:

1. Insert tape or have the tape at the right position in the sequence from start to finish.
2. Plug the microphone into the appropriate jack on the machine.
3. Push the record and start (forward) buttons simultaneously.
4. Continue recording until finished or the tape runs to the end.
5. Push the stop button.
6. Rewind (if directed) to the start of the recording and listen to be sure the recording is adequate.

7. See the teacher or the program for the next step.
8. Remove the tape and return it to the designated location.
9. When finished return the machine to the designated location.

The teacher should follow the same directions as the students with respect to playing and recording. Duplicates of all tapes produced should be kept on file by the teacher in case more copies need to be made. A separate cabinet should be used for this purpose. The duplicate tapes are made on machines that can make four or more copies up to one hour in length in less than ten minutes. Tapes can then be made to correspond to the demand.

A regular maintenance schedule should be followed with the recorders for which the teacher is responsible. The machines used in the carrels should be checked on a weekly basis or more often to be sure that they are playing at a satisfactory level. The cassette compartment should be checked for lint or pieces of foreign material. If any is found, it should be removed. The playing and recording head should be demagnetized and a special head cleaning tape run through the machine at least once each month. A general inspection of the appearance of the machine will indicate severe usage. If headphones are used, the jacks, wire connectors, and the headphones should be checked for appropriate operational levels.

REEL-TO-REEL RECORDERS

These machines are now being used in more sophisticated productions. The quality level is higher than on cassette recorders since the speed of the tape is faster. Recordings made on these tapes may be transferred to cassette tapes at a later time if the teacher desires to use a cassette recorder. The use of this machine for the production of tapes to be used in science facilities is not recommended. The instructors who will be using this machine should decide what features are most desirable for their programs. If the school has a dial-access information retrieval system, this machine would be used. If so, the school would also have a professional audiovisual member of the staff who would be more qualified to make these selections.

Many schools will have reel-to-reel tape recorders available, and these will have to be used until money is available to purchase the easier to use cassette tape recorders. The primary difference between this machine and the cassette recorder is that the tape must be threaded through the reel-to-reel machine with the slick side of the tape toward the front of the machine and then the end must be attached to the take-up reel. Students should use these machines only when they are being supervised or after demonstrating their ability to operate the machine satisfactorily.

TELEVISION EQUIPMENT

No attempt will be made in this section to advise on the type of television equipment that is available or should be purchased. Prices of this equipment are in a continual change. The equipment is being improved, and the prices are dropping. The teacher should be aware that these changes are taking place as well as the implications that this may have on his curriculum.

Portability is a big plus for the new equipment. It is possible to take the camera and recorder out into the field and bring back the pictures of events or physical settings that will stimulate the classroom discussion. Color television tapes and equipment are sufficiently inexpensive to permit any school to afford them. The tapes are now being sold in cassette packages, which eliminates the threading problems of the past. If a purchase is being seriously considered, a salesman of a particular brand should be asked to demonstrate his equipment and the evaluation of each brand should be made with respect to the functions it is to serve. Expertise beyond that of the typical teacher should be present before large orders for equipment are placed.

When a school makes a commitment to instructional television, it will also provide the resource personnel who will assist the teacher in making the maximum use of this tool. These individuals will be available for production as well as maintenance duties. The science teacher would probably be able to become well versed in the use of the equipment because of the number of applications it has for chemistry and physics. Large-scale experiments can be carried out and stored for future reference. Effective use of instructional television equipment will require imagination on the part of the teacher. Since students will seldom be asked to operate this equipment, the teacher need not be concerned with providing them with operating directions.

CAMERAS

Photography can play a large part in illustrating complex problems. To do this work, access to a good camera is necessary. The teacher will frequently use his own as the school's camera may not be available. Among the decisions to make when purchasing a camera are film size, single lens versus twin lens, color versus black and white, type of lens, still versus motion, light meters, and other accessories.

The basic camera in use today is the 35mm camera with options. If the teacher has only the money to purchase one camera, this should be the one that is acquired. With it the teacher can make color and black and white prints or slides for use with the other available projection equipment. The primary limitation of this camera is in making high quality prints. The larger sizes such as

2¼ by 2¼, 3 by 4, and 4 by 5 will provide increasingly higher quality prints. The cost of such prints is also correspondingly higher.

The 35mm camera will, in the majority of cases, be a single lens reflex camera. Basically this means that the focusing and centering is done through the lens that will be used to take the picture. This lens will have a focal length of approximately 50mm, which is a standard length. The lens should also have a variable aperture and a shutter that provides a wide range of shutter speeds for versatility under varying light conditions. The twin lens is typically found on the 2¼ by 2¼ camera, which is used extensively for black and white work. The teacher must be aware that when the edges of the negative are being crowded with the detail of the objects, looking through the lens will not mean that what is seen will necessarily be reproduced on the film.

The decision to use black and white or color film does not determine which camera size to buy as any camera will use either. Slides in color are cheapest in the 35mm format averaging about $.20 each if they are commercially processed, and $.10 or less if processed by the teacher or school using one of the commercially available kits and the film is purchased in 100 feet rolls.

The 35mm cameras can use a variety of lenses with different focal lengths that will allow for close-up wide angle work to the telephoto lenses. A lens with a focal length of 20-35mm will be adequate to handle most wide angle or close-up shots, and one with a 200mm focal length will be adequate for the telephoto work done by most amateurs. The longer focal lengths require a steady hand that eventually requires the use of a tripod for best results. Beyond this length, browsing at the various camera shops will help the teacher determine which lens can do the required job. It is important to select a camera that can accommodate several brands of lenses so that there is some selection in price and quality for the purchaser.

The still- versus motion-pictures decision will depend on what is to be photographed. The stills are cheaper and easier to produce whereas the motion pictures will provide another dimension to the presentation. The 8mm movie camera is the most common type, and some are equipped with zoom lens that allow focusing from an inch to infinity. This will give the photographer greater versatility in photographing many situations. When the teacher becomes proficient he may want to produce his own 8mm film loops for use in his class. It is recommended that this attempt be delayed until skill is developed with still photography unless the school has a production department that can assist the teacher in producing the films. The cost factor in the initial purchase is one difference in deciding which type of photography; the 35mm cameras can start at $50 whereas a good 8mm outfit will start at $100 and upward. Processing is also higher for 8mm films, and the amount of film per sequence is higher as the teacher very rarely utilizes 100 percent of the film shot.

Many of the 35mm cameras have a light-meter attachment that allows the teacher to dial in the film speed as well as shutter speed and then, by adjusting the aperture of the lens, the correct exposure for the given situation is obtained. This is a convenience, and when the meter operates through the lens the success ratio can be over 90 percent. A separate meter is used by professionals to work on effects of varying the lighting and getting the desired exposure level. This option is worth the extra cost in saving shots that might otherwise be lost.

Other accessories are available that may or may not be of use depending upon the situation. Among these are a carrying case, an extended shutter release, and lens filters.

The directions for the use of each camera will come with the camera and should be studied thoroughly before use. Many books are available on photography and effects that should be studied as the skills are acquired through use. Students may be used in class projects in preparing laboratory setups that can be photographed for future use. These students should be under the direction of the teacher when they are using such equipment unless they have previously proved to be capable.

SCREENS

The type of screen that is used to show films or for projecting images does not vary a great deal. The main criteria is the amount of light that is actually reflected. The glass beaded screens do a satisfactory job, and the difference between this and other screens is hard to distinguish without a meter. Size is the other variable. Large rooms require large screens so the pictures may be enlarged to provide a maximum viewing area. The sizes then progress downward to the small screens used in the carrels for individual showings.

These screens should be permanently mounted in each classroom and pulled down when needed. A special mounting for a screen that is used with an overhead projector is advantageous. This screen should be sloped out into the room from its base to eliminate much of the distortion that accompanies the tilted head on the projector. The screens used in carrels are small and usually mounted in one place, which serves for all type of projectors used there. Portable screens can be used when no other type is available, but should be the exception rather than the rule.

CALCULATORS

Electronic calculators are becoming a part of many science classrooms. They eliminate much of the tedious work or mathematics and allow the students to concentrate their time upon solving problems. Prices on these machines begin at $100 and go up depending upon the features requested. A bare minimum would include the functions of multiplication, division, addition, subtraction, and square root. In addition, the calculator should be equipped with two memory banks for complex calculations. Other functions that may be added include trig functions, $\log_{10}$, $\log_e$, and the like. Calculators in the $5,000 range are programmable to use functions that may be unique to the individual school's situation. These calculators should be entirely electronic with a display board for the answers. This eliminates many of the maintenance problems associated with mechanical calculators.

Students can use these calculators after a very brief period of instruction. The instruction booklets that accompany these machines are usually sufficient for this purpose. Copies of the instructions should be obtained so that an adequate supply is maintained and that loss will not be detrimental. The student has to be trying if damage is done to these machines. Students should have access to these machines both during and after class. They will usually do more problems in a shorter period of time when a calculator is available to them.

Instructional Design and the Use of Audiovisual Materials

After deciding that various audiovisuals are a desirable part of the instructional process, the teacher must make them an integral part of the instruction. An examination of the development of instruction will indicate the need for these audiovisual materials and how their effects may be evaluated. It should be realized that a commitment to use audiovisual materials in a group presentation is different from providing different modes of input by means of individualized instruction.

The procedure that has been developed for writing programmed instruction can be applied to all levels of instruction. The basic premise is that if the instruction fails to produce the desired behavioral changes, then the program is at fault and should be redesigned. The philosophy and the procedure used in program development can be extended for daily classroom instruction as well. (The details of programming are spelled out in Chapter 10, Programmed and Computerized Instruction from *Toward More Effective Science Instruction In Secondary Science Instruction in Science Education* by Andersen and Koutnik, New York: The Macmillan Company, 1972). The basic procedure is as follows:

1. Selection of the content to be learned.
2. Statement of the specific behavioral objectives.

3. Preparation of the initial rough draft.
4. Revision after the appropriate feedback.
5. Testing.

The strategy for the classroom is developed in step 3. The teacher has many options in the presentation of material. It may come from a lecture, a laboratory, a short film, or other sources. The selection depends in part upon the teacher's assessment of the student interest level for a given mode of presentation and the ease with which the material is used when it is presented in that mode. Ultimately the decision will rest on the intuitive feeling that the teacher will have when the method used, the content, and the students are compared and contrasted. The problem for the beginning teacher is that this intuitive feeling will not always be present. If he chooses wrongly, the entire class may lose a day. This loss is hard to justify on a continuing basis, and if the students were not flexible, it could not be tolerated at all. Usually any method will have some success with the students, and the teacher should write out an evaluation of the apparent success or failure and the factors that effected that situation. Ideally the teacher could test for objective mastery, but this is not practical on a daily basis. If the student cannot get the input from the classroom or text that will enable him to perform successfully on the test, the opportunity to learn that material or capability is lost for that student. The teacher must choose the strategy wisely if all students are to master the materials.

When the content associated with a given set of objectives is presented to the student in more than one way, and the student has his choice of information input, then the probability of success and hence the motivation level should be increased. The implication is that the student must have a choice, which means that a commitment to individualized instruction has been made. The implementation of individualized instruction is as varied as the teachers who are doing it. Every teacher sees his situation as being slightly different and makes accommodations that insure the student's success. The student now has time and alternate routes to success that were not available before. Most of these efforts to individualize the instruction rely entirely on the use of the audiovisual materials.

Individualized instruction should be used in the chemistry or physics classroom in conjunction with regular class work. Remedial material should be available at the first of the year if the students do not possess all of the prerequisite skills that are necessary to insure success. Enrichment or tangential material can be made available through the year for those who desire independent expansion. Part of the tedious problem-solving skills can be learned best by a well constructed program thus freeing the teacher of a boring, repetitious job and allow him to work individually with students having problems. To fulfill these objectives, learning carrels should be available in the classroom, laboratory, or nearby classroom. These carrels should contain all the essentials needed for any program and would include the projectors, recorders, and software. If certain laboratory work is to be done individually, then a sink, water, gas, and electricity should also be provided. It is the nature of most carrels to be small compartments; when laboratory work in chemistry is done this way the students tend to be out of sight of the instructor, which is not desirable with respect to safety. These carrels should also be available to the students at all times without disrupting any other classes that may be using that room. If the students are to be allowed to work at their own pace, time must be provided to allow for the repetition of some material.

The teacher who desires to use individualized instruction must choose the carrel's material. He obviously has the choice of designing his own material or using that available from commercial sources. Whatever its source, the material in the carrels should be examined with the following in mind:

1. What is the program designed to teach?
2. What audience was it designed to reach?
3. Under what conditions will the program be taught?
4. What evidence is presented that it is successful?
5. How much student time is required to attain the stated objectives?

The program that is prepared or selected must have a high probability of teaching the students those concepts or skills that the teacher wants the students to acquire. To justify the use of the program it must demonstrate, by some criteria, that it is capable of accomplishing what it says it can do. The best programs will provide data indicating the nature of the student population upon which it was validated. It will indicate entry level performance, final performance, and net gains by the students on which it was tested. From this information the teacher can decide if the program will be successful with his students. The amount of time that is necessary to reach the given level of performance is also important because it is not desirable to have the students spend an excessive amount of time on a program when other shorter formats can change the students' behavior to the same extent.

The teacher-prepared program should undergo the same analysis. It should be able to demonstrate the capability of changing the students' behavior in line with the stated objectives. If it does not, the program should be revised and retested. For this reason teacher-prepared materials should not be tested on whole classes until the teacher gets an opportunity to test the program on a selected number of students individually. Most of the major weaknesses of the program should be eliminated before a whole class is asked to work through the program. The teacher must be careful in selecting students for the initial trial of the program as a few students would not give an accurate picture of the way the entire student population would respond. Picking the brightest students will not indicate how the slower students might respond to it. The student population to be reached by the program must be clearly defined and the testing must then be done with a representative sample of that group.

Material Required for the Preparation of Audiovisual Materials

The following lists of equipment and supplies are included to indicate what materials will be needed to make the various types of audiovisuals that have already been discussed:

2 by 2 slides
1. A 35mm camera
2. An exposure meter
3. Film (color and/or black and white)
4. Floodlights
5. A copying stand with lights
6. A tripod
7. Filing boxes
8. A slide sorter/viewer
9. Cardboard mounts
10. Tanks and chemicals for development of the film

Darkroom equipment
1. Enlarger handling negative sizes from 35mm to 4" by 5"
2. Contact printer
3. Development chemicals for film and paper
4. Photographic paper
5. Developing tanks for film
6. Developing trays for photographic paper
7. Thermometers
8. A timer with a time range of 1 second to 10 minutes
9. A timer for the enlarger to regulate the exposure
10. Safe lights
11. A sink with hot and cold running water
12. Funnels
13. Drying racks
14. A measuring cup graduated in cc. and oz. to 1 quart
15. Sponges

16. Assorted bottles
17. A negative brush
18. Slide drying racks
19. A print washer
20. A paper safe

Transparencies
1. A dry mount press
2. A tacking iron
3. A wet-process transparency maker
4. A dry-process transparency maker
5. Assorted grease pencils

Miscellaneous
1. A Leroy lettering set
2. A drafting table
3. Stencils of various sizes
4. Metal rulers
5. A paper cutter
6. India ink
7. Laminating film
8. A clipboard
9. Grommet set plus grommets
10. A light box

REFERENCES
General Audio-visual Production and Utilization Books

Brown, James W., Richard B. Lewis, and Fred F. Harcleroad, *A-V Instruction, Materials and Methods*, 2d ed. (New York: McGraw-Hill Book Company, Inc., 1964).

Dale, Edgar, *Audio-visual Methods in Teaching*, 3rd ed. (New York: Holt, Rinehart and Winston, Inc., 1969).

Erickson, Carlton, W. H., and David H. Curl, *Fundamentals of Teaching with Audiovisual Technology*, 2/e (New York: The MacMillan Company, 1972).

Frye, Harvey R., and Ed Minor, ed., *Techniques for Producing Visual Instructional Media* (New York: McGraw-Hill Book Company, Inc., 1970).

DeKieffer, Robert, and Lee W. Cochran, *Manual of Audio-Visual Techniques*, 2d ed. (Englewood Cliffs, N.J.: Prentice-Hall, Inc., 1961).

Mager, Robert F., *Preparing Instructional Objectives* (San Francisco: Fearon Publishers, Inc., 1962).

Special Topic Production Books

Bryce, Mayo J., and Harry B. Green, *Teacher's Craft Manual* (San Francisco: Fearon Publishers, Inc., 1956).

Coffelt, Kenneth, *Basic Design and Utilization of Instructional Television*, Visual Instruction Bureau, University of Texas, Austin, Texas.

Coltharp, Joe, *Production of 2 x 2 Inch Slides for School Use*, Visual Instruction Bureau, University of Texas, Austin, Texas.

East, Marjorie and Edgar Dale, *Display for Learning* (New York: Holt, Rinehart and Winston, Inc., 1952).

Edling, Jack V., et al., *Four Case Studies in Programmed Instruction*, Fund for the Advancement of Education, 1964.

Frye, Edward B., *Teaching Machines and Programmed Learning: An Introduction to Autoinstruction* (New York: McGraw-Hill Book Company, Inc., 1962).

Frye, Roy A., *Using Tearsheets*, Visual Instruction Bureau, University of Texas.

Green, Edward J., *The Learning Process and Programmed Instruction* (New York: Holt, Rinehart and Winston, Inc., 1962).

Guimarin, Spencer, *Lettering Techniques*, Visual Instruction Bureau, University of Texas, Austin, Texas.

Linker, Jerry M., *Designing Instructional Visuals: Theory, Composition, Implementation*, Visual Instruction Bureau, University of Texas, Austin, Texas, 1968.

Linker, Jerry M., *Instructional Display Boards*, Instructional Media Center, University of Texas, Austin, Texas.

Lockridge, Preston J., *Better Bulletin Board Displays*, Visual Instruction Bureau, University of Texas, Austin, Texas.

Lockridge, Preston J., *Educational Displays and Exhibits*, Visual Instruction Bureau, University of Texas, Austin, Texas.

Meeks, Martha F., *Models for Teaching*, Visual Instruction Bureau, University of Texas, Austin, Texas.

National Society for the Study of Education, *Programmed Instruction*, Sixty-sixth Yearbook, Chicago: University of Chicago Press, 1967.

Pipe, Peter, *Practical Programming* (New York: Holt, Rinehart and Winston, Inc., 1966).

Sloan, Robert, Jr., *The Tape Recorder*, Instructional Media Center, University of Texas, Austin, Texas.

Smith, Richard E., *Local Production Techniques*, Instructional Media Center, University of Texas, Austin, Texas.

Smith, Richard E., *The Overhead System: Production Implementation and Utilization*, Visual Instruction Bureau, University of Texas, Austin, Texas, 1966.

Free and Inexpensive Materials

The science teacher is usually committed to using all of the available resources at his disposal. He first compiles the chemicals and equipment in his laboratory and proceeds from there to add audiovisual and demonstration equipment to the extent allowed by the school budget. School budgets limit the amount of money that an individual teacher can spend, but the resourceful teacher will be able to acquire as much material for his money as possible.

Audiovisual materials may be secured from manufacturers in the desired field. Processes are detailed in films, charts, and descriptive pamphlets. All of this is either free or available for free loan. The teacher must, however, be willing to accept the advertising that accompanies the pamphlets. The teacher should explore the availability of free or loan materials from these sources before buying or renting comparable material. The addresses of many of these companies may be found in the Appendix.

Materials are usually difficult to acquire for no cost. Occasionally, if the teacher has established a good relationship with local companies, he may acquire suitable equipment when it is no longer needed by a company. The equipment available from this source will usually be older but still in operable condition. Even damaged equipment may be used as a source of parts from which a variety of equipment may be constructed. This is particularly true of electronics equipment. Discarded television sets may be acquired from local dealers who will usually not want to spend the money necessary to salvage the tubes, resistors, and transformers that may still be usable.

The individual teacher can locate a considerable amount of free audiovisual material if he is willing to write many letters. Fortunately, several publications are designed to inform the teacher of the availability of free material. The two publications consulted most frequently by science instructors are *Educators Guide to Free Science Materials* and *Educators Guide to Free Films.* These two publications are updated annually and provide the teacher with a variety of free resources.

The *Educators Guide to Free Science Materials* is available from Educators Progress Service, Inc., Randolph, Wisconsin 53956 at a price of $9.25. The areas covered include films, filmstrips, tapes and transcriptions, charts, exhibits, magazines, and posters, and a variety of other printed materials. Each item is reviewed for the teacher, which will help the teacher decide if the material would be a valuable addition to the classroom. Helpful hints are also included on the time between order and delivery for film loan orders and who is expected to pay the return postage. Although *Educators Guide to Free Science Materials* covers all aspects of science from elementary levels through high school, chemistry and physics teachers will find a considerable amount of useful material.

The *Educators Guide to Free Films* is also available from Educators Progress Service, Inc., Randolph, Wisconsin 53956. This publication is updated annually to prevent the teacher from ordering films that are no longer available. Films from a large variety of sources dealing with the complete spectrum of topics are reviewed. Detailed ordering instructions are included. All science films are included in both of these publications. This publication is not as valuable to the science teacher as some of the other publication, but it should be made available to all teachers in the school system. The teacher will usually order enough free material to more than cover the cost of either publication.

The teacher may also consult current magazines for the most recent material. Manufacturers frequently advertise brochures in *Scientific American, Journal of Chemical Education,* and *Physics Today.* Some publications provide request cards for the reader to mail for more information. Single copies of the information requested will usually be provided. The teacher may want to write directly to a company offering a new product or process that could be interesting to a class. Material is often available even if it has not been advertised. The teacher should observe the guidelines for letter writing when requesting any information.

Free equipment is naturally harder to acquire, and most of it will be used or obsolete. Sources to check for free equipment include local manufacturers, electronic repair shops, government surplus, and other educational institutions.

The teacher must be selective in requesting material from local manufacturers. There is a fine line between asking a company to donate equipment for public relations value and outright begging. Electronic manufacturers will often donate new equipment if the school is offering evening courses that would be available to their employees. The teacher should keep the principal and/or school superintendant informed of his activities involving requests for equipment. They can frequently assist the teacher by providing the names of the appropriate individuals for the teacher to contact. They may want the teacher to refrain from requesting from some companies so that other cooperative projects between the business and the school are not jeopardized. Teacher contact should be with the companies nearest to the school and preferably within the school district boundaries. From this point the teacher can establish relations through the public relations department of companies within the city, local area, or possibly within the state. Working with companies that are close to the school will usually provide the best results.

Government surplus material is often overlooked by many teachers. Each state has an outlet that makes materials and supplies available to local governmental units. Business managers are aware of their existence but do not know what the teacher needs specifically. The physics teacher can acquire gears, electronic equipment, hydraulic equipment, motors, and other odds and ends that can be used to make equipment. Outdated oscilloscopes are occasionally also available. The source of these materials is usually government laboratories and surplus from the armed services. Frequently the surplus equipment has never been used. If one of these outlets is near the school district, the teacher should make arrangements through the school to visit these centers. Requests for individual types of equipment can also be made that will be honored as it becomes available. The price on this equipment is usually less than 10 percent of its original value. The chemist will find some chemicals and a variety of glassware. Periodic checks on this source will frequently uncover a piece of equipment that will be valuable to the science program.

Other public institutions with laboratories should be contacted and their possibilities recorded for future reference. When new equipment is purchased, the old will either be scrapped, traded in, or kept in operation. If it is salvageable, the teacher may want to have the equipment repaired for use in his program. Frequently these laboratories are not aware that there is another local institution that could use their old equipment. Hence, the teacher does not know that it exists. Places that the teacher should visit and contact include hospitals, water treatment plants, waste treatment plants, and public health laboratories.

Salesmen from equipment supply houses will also know when they have received large orders and will be getting trade-ins. The teacher can either buy the trade-in as is or after the supply house has reconditioned it for less than the original price. Salesmen are also good sources of information on who, what, when, and where there will be phasing out of old equipment. Contact should be made with those individuals or institutions as soon as possible.

Acquisition of free materials will begin with a letter to the appropriate company. The teacher should take special care with the letter because of the impression it creates. The teacher will probably be dealing with the company again in the future and certain rules should be followed to build a good working relationship that will provide the teacher with necessary materials. The teacher should be particularly careful with the following points:

1. Use school letterhead stationery.
2. Keep the letter short.
3. State exactly what is wanted.
4. Request by topic.
5. Keep requests reasonable.
6. Follow instructions.
7. Recheck addresses and titles.
8. Follow up with a thank you letter.

The companies that have been selected as sources of material will be receiving many requests for the information sought by the teacher. The companies are under no obligation to fill every request, and the neatest and easiest-to-handle requests will receive preferential treatment. Odd size paper does not file easily so standardized stationery should be used. The school letterhead stationery will also indicate that the request is a genuine one from a teacher insuring to some extent that the materials will be used in the classroom. The school secretary can be asked to type the letter if the teacher cannot type. Spelling should be checked on the rough draft and again before the final letter is sent. The teacher's position should also be included following his name, and if he has another title, that should be used as well. This enables the company to better evaluate the request. Films are not typically loaned to an individual without some assurance of their return. The letterhead stationery along with the position held by the teacher provide the companies the confidence to honor the requests.

Keep the letter short. The company will receive many requests and the letter should appear as if it would be easy to answer. Only the essentials of what is being requested and why should be included. It should be necessary to use no more than two paragraphs to accomplish this objective.

State exactly what is being requested. Order numbers should be used if known. When materials are advertised in magazines, the company will also have specific box numbers to write to as well as special codes for the material being distributed. Mentioning the source of the advertisement will help the company measure the effectiveness of its advertising in that publication.

The teacher should make his requests by topics. If the teacher has not seen a particular advertisement or does not know of a specific offering, then his requests should be very detailed. The objective for the material should also be carefully explained so that the company can match its materials with the needs of the teacher. The company will dislike and perhaps disregard all-inclusive requests. A request for information about the production of certain chemicals from a chemical company is preferred to a request for general information about what the company produces.

Requests should be for a reasonable quantity of materials. If the teacher has no indication of the availability of material from that source, his requests should be kept to one copy of the material until an evaluation of that material has been made. Not all of the available aids will be suitable for use in the teacher's school. Other schools could possibly use those aids, and if the supply of the company has been exhausted, those schools that could use them will not be able to. When the teacher has made an evaluation of the material, he should then request larger quantities. Even if the audiovisual materials are free to the teacher, they are not free to the company, and this fact should be recognized and respected.

Follow the instructions when ordering. Most ads will give set procedures that should be followed to receive the materials in the shortest period of time. The company will receive many requests and will honor those that are easiest to handle.

A brief statement of intended use of the materials will punctuate the request. This should always be included when ordering large quantities. An indication that the material has been reviewed and found acceptable will facilitate the acquisition of larger quantities.

Before sending the letter, the teacher should check to be sure that all addresses are correct. This includes the company address and the return address. If possible, a self-addressed shipping label could be included that will save the company time and money, and should eliminate incorrect addresses. Titles of the persons from whom the teaching aids are being requested and the teacher's title should be included and correct.

When requests have been honored, there should be a following thank you letter indicating that the materials were received and used. Such attention to detail will create a favorable impression and could lead to future offers of material that might not be made to the general public. This letter should also be kept short and to the point.

Once the materials have been requested, the teacher must allow a reasonable amount of time for delivery. The usual minimum is six weeks and occasionally longer. No concrete plans for the use of the materials should be formulated until they actually arrive.

Films requested that must be returned will follow the same pattern. The first request for films might best be used for review purposes. If the class time is available after review and the film is valuable at that point in the instructional sequence, it should be used. If not, plans should be made to schedule a showing of the film the following year close to the time when it would be most beneficial to the students. The request for the following year should be made as soon as possible after the decision to use the film has been made. This will increase the probability of the film's being available. All such letters should include at least one alternate date and preferable two. If the film can only be used once, the letter should indicate this fact so that the film may be used by other interested parties.

The following letter is included as a sample that may be copied or modified and includes the essentials for requesting teaching materials that have been discussed.

Ford River Public Schools
Ford River, Michigan 49807

Superintendent   Phone

October 31, 1975

Mr. David Edwards
Public Relations Manager
Excellent Chemical Company
Capitol City, New York 11016

Dear Mr. Edwards:

I read of your offer in the March issue of *Scientific American* and would appreciate receiving one copy of:

Removing Mercury from Industrial Wastes.

I will be using this booklet in conjunction with our study of mercury in my chemistry class.

Thank you for your assistance.

Sincerely,

Patrick R. Hittle
Chemistry Teacher

jsl

When the supply of free materials has been exhausted temporarily, for whatever reasons, the teacher should consider other possibilities. Can the desired piece of equipment be constructed or can a reasonable substitute for it be found? A large number of projects to build are described in the issues of many of the electronics magazines on the market. *Scientific American* has a monthly feature on building some piece of equipment that can be used to investigate various physical phenomena. The parts are often already available; and if the teacher is willing to spend some time following directions, a useful piece of equipment will result. Other ideas can be found in hobby books that are available from the public or school libraries. Another source of ideas can be obtained from colleagues. Letters outlining a specific problem could be sent to area schools to determine if they have had the same need and if so how it was solved. If the teacher remembers that all complex machinery and equipment is just a collection of the simple machines, an effective substitute can usually be found that will demonstrate the given principle.

The Functional Laboratory

The choice of the traditional laboratory portion of the high school chemistry and physics course or the inquiry-investigative approach of the new programs presents one of the greater challenges to the neophyte teacher. It is here in the laboratory that the student can become involved with his environment in such a way that the theories and principles encountered and discussed in the content or lecture portion of the course acquire physical meaning.

The challenge of science teaching found in either approach is to involve the student to the maximum extent possible with the concepts of the field of study. What is the best way to accomplish this objective? What activities should be used? How and when should the students be involved? What are the hazards involved with each type of activity? These and other questions confront every high school chemistry and physics teacher. The answers to these questions vary as would be expected, but the environmental emphasis now being stressed in education is causing a shift in thinking and actual practice. This shift is designed to involve the students more directly in answering relevant questions.

To participate actively in the search for knowledge, the student must have certain skills with respect to the acquisition and the utilization of data. Data collecting in and of itself will serve no useful function. Knowing what data to collect is far more important, but the student cannot know what data to collect until he has gained the skill of proposing the questions that are to be answered. The questions in turn cannot be proposed until the student is able to evaluate the various components of the environment and critically analyze them in terms of what information will be needed to answer the questions. Having collected the data, the student must then interpret it in the correct manner and evaluate that data with respect to the methods used in its collection before he can accept or reject his initial solution to the problem.

The laboratory must function to provide the student with the required skills for the generation of knowledge. These skills are vital to the student because of new problems he will be asked to answer and areas of knowledge that will come into existence during his lifetime. It will also provide answers to traditional problems that have been accepted as worthy of answering. The acquisition of knowledge-generation skills must stand as the foremost objective of the laboratory portion of the high school chemistry course.

The new programs utilizing the inquiry approach have developed extensive materials that will assist the teacher in reaching the stated objectives with respect to the laboratory work. These procedures will not be repeated here. Instead, teachers are urged to examine each program's (PSSC, HPP, CHEMS, and CBA) material and apply it as it fits the individual's situation. The remaining portion of this chapter will be directed to making changes in the existing frameworks.

The high school chemistry course has traditionally been segmented into a lecture portion and a laboratory portion. The lecture has been used primarily as a means of information dissemination. Students have been expected to take notes on the material "covered" in class and to reproduce that material on future tests. Very little opportunity has been given to the frank and open discussion of problems that require chemical knowledge and research to arrive at viable solutions.

The laboratory portion of the course was initially designed to provide "experiences" whereby the student could come to understand what the research chemist does. Typically the student starts with a complete set of instructions that inform him of the exact procedures he will use and observations he will make during the laboratory period. He is then provided with questions that require him to fill in the appropriate word in the blank. It is unfortunate that the world does not provide such questions for the research chemist or the scientist in general to answer.

The function of dispensing information by the teacher is being replaced. The teacher is becoming a designer of environmental encounters. Some psychologists contend that a student does not truly acquire information that is structured in his mind until he has had the opportunity to physically act upon it. It would follow then that the more energy the student expends acting on gathering and processing information the greater would be the probability that it would be incorporated into his cognitive framework in some meaningful fashion. Thus, it should be the role of the teacher to provide as many opportunities as possible for the student to become involved in collecting and organizing information. Students can read faster than the average teacher can speak. It would seem to be advantageous to assign reading that provides the best presentation of the background material the student will need to act on a given problem situation.

Having assigned the information-dispensing responsibility to textbooks or carrel experiences, the teacher must now decide what adequately written objectives will be accomplished in the remaining time.

Since science is a problem-solving activity, the time could be spent introducing the students to problem-solving procedures. Problems for consideration may be presented by demonstration, films or film-loops, or by laboratory activities in which the student defines the problem to be studied and the method for studying that problem.

The material that the chemistry or physics teacher presents in class depends to a great extent upon the availability of support materials. Selection and procurement of books, programs, film loops, cassette programs, and the like will determine which objectives must be achieved by classwork. Ideally, a group session can stress the difficulties that the students had with particular problems or content material. It is hoped that eventually all of the class time can be spent in problem-solving activities either in the generation of the problems or in determining possible solutions.

In the secondary schools there is a large probability that some teachers will still be confronted with the traditional lecture-laboratory program. As a first-year teacher, it is not always possible to institute all of the changes that might be desirable. It is, therefore, imperative that some plan of action be formulated to cover the first year's activities until more appropriate procedures can be implemented. It is recommended that the teacher study the textbook and library resource materials to determine what content may be covered with homework assignments and what concepts or cognitive skills must be developed by classroom activity. With this background the teacher can sequence the activities that will achieve the stated objectives.

A list of the objectives for the course or at least for the first six weeks will aid the teacher in deciding how the available resources may be used to maximize the learning of the students. As performance objectives have already been treated in this book, the reader is urged to consult the many examples already given and to use or modify those that would seem appropriate to the course of study being planned.

For the nonlaboratory part of the course, the beginning teacher is urged to use an inquiry approach to the accomplishment of all classroom objectives. Inquiry is here defined to be the learning activity exercised by the student in his attempt to relate the new concepts to those cognitive frameworks he already possesses. A more complete understanding of inquiry may be had by consulting the bibliography at the end of this chapter.

To accomplish the objective of increased problem-solving abilities of the students with respect to chemistry or physics, the teacher has many options. It is possible to demonstrate "type" problems; and by giving the rule and following this with examples and homework, some of the students will learn to solve that particular type of problem. The student may also be assigned reading materials that present the rule and work out various examples along with suggested problems for him to try. The teacher then has the option of breaking the class into smaller groups and proposing various problems for each group to answer based on the material that was to have been studied outside of class. In groups of four to six, students can work cooperatively and those who have difficulties can then question the other members of the group. Ideally everyone should understand the particular "type" of problem under consideration. After a certain number of different types of problems have been mastered, a more complex problem should be proposed, either directly or indirectly, to the class so that the skill of vertial problem-solving may be developed by each student.

The quantity of time devoted to the laboratory portion of the course will depend on: (a) the syllabus for the course, (b) other teachers giving the same course, (c) the departmental chairman, or (d) the teacher himself. In lieu of other guide lines, it is a common practice to devote approximately 40 percent of the class time to laboratory activities. This may be distributed on a fixed, twice a week schedule, or it may be varied to fit the sequence of instruction.

Selecting Experiments

The selection of an experiment should depend upon:

a. the quality of the concept to be illustrated.
b. the number of observations possible.
c. flexibility of procedure.
d. skills to be acquired.
e. simplicity with respect to equipment and chemicals.
f. the probability of completion without accident.

The quality of an experiment may be measured in many ways. To some teachers there are experiments that are classics, that is, they are traditionally in the curriculum and should therefore be continued. If one of the objectives of the chemistry or physics course is for the student to become aware of his environment from a particular point of view, then a quality experiment will be one that illustrates some basic concept in a context that can be related to some aspect of the environmental problem. For example, purification of water and/or properties of water take on increased importance with respect to the quantity of pure drinking water needed and the limited sources available to all municipalities. Solution chemistry takes on added importance in terms of waste disposal because wastes are presently being dumped into the air, lakes and streams, or into the ground. The quality of an experiment must also be considered with respect to the questions it can generate and the solutions that may be proposed as a result of the completed work. It is imperative that an experiment be one that will challenge the student to become involved.

Some experiments can be performed with the objective that the students make one key observation. However, one must examine the purpose served by those experiments when compared to an experiment in which the students can make many functional observations. When many observations are possible, the student must acquire a skill in discriminating among the observations and selecting those that directly pertain to the problem being considered. It is recommended that when a teacher selects an experiment he should actually perform the entire experiment himself or possibly with one student before assigning that experiment to the entire group. From this experience the teacher can determine the difficulties the students will encounter and can have data available to check the data that the students will obtain later.

An experiment should be open-ended. If the answers to the questions proposed by an experiment can be answered without actually doing the work in the laboratory, the teacher should be able to justify using that experiment. An open-ended experiment is probably best characterized by the fact the answers to the questions are not known in advance. Quite often the questions to be answered must be generated by the students. The experiments written by the Chemical Bond Approach Project illustrates one of the better attempts to construct open-ended experiments. It can be noted that this type of experiment will produce a number of procedures for collecting data that will help to answer the proposed question.

The laboratory skills that each experiment will develop should be given consideration in the selection of all experiments, although it should not be the primary consideration. It must be considered that the majority of the students in a high school chemistry and physics class will not go on to major in either subject in college. The only skills developed should be those that will be used often during the year they are studying chemistry or physics. Colleges have changed the emphasis in their laboratory programs for chemistry and physics majors. This change has resulted in data collection from electronic instruments rather than from "wet lab" procedures. For these and other reasons the students should not be evaluated extensively with respect to his laboratory skills.

Developing an Inquiry Laboratory Experience

The success or failure of the inquiry approach to chemistry or physics instruction will depend on the willingness of the teacher to spend time in developing ideas for the class to pursue. Many good ideas are available in the commercially prepared laboratory manuals which should not be overlooked by the teacher. The following scheme is presented as a means of taking the better ideas and transforming them into a usable inquiry approach.

Inquiry can also be defined as the active approach to chemistry or answers to questions they have proposed. To get these answers the student must be encouraged to pursue many different means to obtain the necessary data. Some of this data may be collected in the laboratory as a group or smaller groups within the class may be called upon to obtain different data that can be pooled at a later time. Other sources of data such as books and magazines should also be considered to enable the students to formulate and answer as many questions as possible during the year they are in the chemistry or physics class.

The teacher may be able to start with student questions the very first day and develop the daily objectives from that point. This is not usually possible. Instead the teacher should select at least one broad area to serve as the entry point onto the subject. Two areas that should be considered for an entry point are: "What are the possible sources of energy available to man?" and "What can be done to clean up our air and water?" Students may be motivated and challenged by these or other questons that they may pose. At the beginning of the year the teacher can select experimental activities related to these questions from the regular laboratory manuals available.

The regular experiment is characterized by the fill-in-the-blank approach to gathering information. The teacher is expected to extract the basic idea or concept of the experiment and provide a rationale for its relation to the larger question that the teacher may want the students to explore. This idea may then be proposed to the class directly in question form or as a demonstration. Neither the directions or all of the questions that could be answered should be handed out at this time.

Before the teacher presents the idea to the students, a thorough study of the idea should be made by the teacher. This study would include the rationale for its use. A list of questions that are related to this idea should be formulated in advance to serve as a guide to the teacher when the students prepare their own questions on this concept. The questions that may be answered by the laboratory work should be starred and a list of procedures for collecting data

210 Additional Resources for the Teacher of Chemistry and Physics

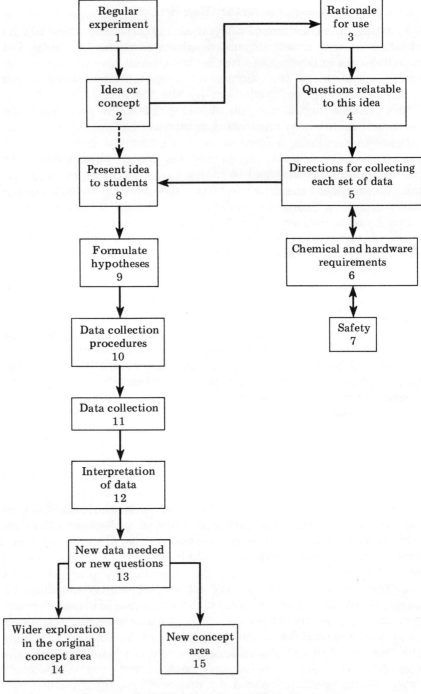

Figure 4.3 Development of an inquiry experience.

on these questions should be prepared. These procedures will insure that when directions are developed all the necessary steps will be included so that unsafe situations will be minimized.* Steps 1-5 should be completed by the teacher before any presentation to the students. Steps 3, 4, and 5 represent the teacher relating to the course objectives.

The presentation of the idea to the students should be accomplished in such a way that the students will be interested and will be able to formualte a large number of questions about the idea. The presentation will be enhanced by the teacher's imagination, and the possibilities include demonstrations such as those outlined earlier, film loops from commercial sources, movies, reading assignments, slides, filmstrips, and so on.

The students are then expected to develop the questions or hypotheses through group discussion. As the year progresses, the students will develop the abilities to note the trivial questions and to concentrate on the more relevant ones. Initially, the teacher may have to make the students consider each question seriously as to its relative value in the content of the overall concept.

Once the questions have been formulated and clarified, data collection procedures will have to be developed. Using this as a class project will enable small groups of students to be formed to explore different problems at the same time and to present their conclusions to the entire class. Those questions that the group, with a little teacher assistance, decides should be answered by collecting data in the laboratory are the ones for which procedures must be developed. The teacher's experience with the questions will determine how much freedom is to be allowed to the students for determining the procedures to be followed. Students may be given free access to the study of circuits in which wires, bulbs, and batteries are used. They should not be allowed to explore what happens when two solutions are mixed unless the teacher has checked all possible combinations. Safety must be considered when the students are allowed to explore in the laboratory without a formal set of directions for the collection of data.

The data collection in the laboratory will occur during the course of a class period in which no other activities are planned. Most fifty-five minute periods will require all of the student's time for setting up the equipment, making the required manipulations, recording the observations, and returning the laboratory to its original condition. The students will then analyze their data in a separate class period. This will give the teacher sufficient time to prepare any

(*Figure 4-3 illustrates the sequence of this procedure for the development of an inquiry laboratory experience.)

chemicals or to obtain any equipment that is not on hand before the class begins.

In the assignment of the questions to groups, some groups will be experimenting actively in the laboratory and others will be considering resources in the library. Not everyone needs to perform the experimental part unless the teacher has determined that the performance of the experiment is a vital and necessary part of the student's interaction with the environment. When the students meet again after collecting the data, a period of class analysis of the data should be conducted. Each group should report to the entire class concerning the information that it has collected. Visual presentation via chalkboards, graphs, charts, and the like will provide greater understanding. The teacher should have a list of the questions generated earlier duplicated and distributed to the class. The handout should have enough space between the questions for the answers that are developed in class.

Not all of the questions formulated by the students are likely to be answered by the students from the sources examined. At this point the class is ready to consider either a new set of questions or determine the other activities need to be conducted to answer the old questions. The teacher should function at this point to evaluate what gains might occur to the student by the continuation of work on the original questions. Another point of reference here is considering the other areas that can and should be explored by the class.

The class, with the assistance of the teacher, is now ready to repeat the process with a new set of questions that will stretch the concept at some boundary point. Another possibility is to leave this concept area and to begin explorations in another area.

The teacher should prepare a chemical and equipment needs list for each experiment that is seriously considered. (An example of one form that may be used is given in the Appendix.) These lists may be used to determine what equipment and which chemicals and their quantities are needed for the complete year. All experiments should be considered in terms of the complexity of equipment needed and the types of chemicals used. Simplicity will usually produce the same quality results as many complex experiments. Unusual properties of all chemicals should be noted for safety considerations when the experiment is actually performed.

Safety must be of primary concern to the science teachers at all times. All possible consequences should be noted when the equipment and chemical needs sheet is filled out. Consideration should then be given to those chemicals and equipment that offer a maximum in all of the other areas and a minimum of potential hazards. In any case the teacher should be well prepared for anything unusual.

The teacher is urged to give special consideration to the chapter on laboratory safety when planning the laboratory program of experiments.

Laboratory Reports

After spending considerable time preparing for work in the laboratory, the teacher faces the problem of determining what the students are to do with respect to reporting on their results. Typically, students are given a sheet of questions or statements that contain blanks to be filled in or space for short answers. This traditional approach is easy for the teacher to grade but does not provide for the acquisition of skills in reporting that the student should possess. If an inquiry approach is used in the laboratory similar to the one described, these short answer sheets are not appropriate so the teacher must plan an alternative course of action.

Operating on the assumption that the students will be developing the problem area and procedures, with the teacher's guidance, it is recommended that the student adhere to the following format as closely as possible.

1. A definition of the problem area.
2. A list of the questions to be answered.
3. A statement of what data will be collected to answer the questions.
4. A description of the procedures to be used to collect each piece of data.
5. A record of the observations and comments on the data collection process.
6. An organized compilation of the data collected.
7. A list of the mathematical formulas used in treating the raw data.
8. Answers to the questions initially posed.
9. An evaluation of the problem and procedures.
10. A list of questions to be answered in future experiments that were generated by the work on this problem.

In the discussion preceeding the actual laboratory work the teacher will have to assist the students in reducing the initial problem to a manageable one. For example, the students may wish to study air pollution and the teacher should help them pinpoint a given segment from which all work can begin. One starting point could be the properties of oxygen that might include the physical and chemical properties of the gas. Although an area of this magnitude could be directed into an entire semester's study, the teacher must help the students select the individual problems in a sequence that will yield the highest probability of success. Once the problem has been clearly defined by the students, they should record it in a notebook for their future reference.

Along with a problem area comes a number of questions associated with that problem. Some of the problems will be so familiar to the teacher as to be trivial. A point can be made for the inclusion of such questions on the grounds that they are questions the student would like to have answered. Lists of such questions should be compiled by each class and recorded in a notebook by the student.

Once the questions have been asked, information will have to be gathered to assist in answering those questions. At this point the teacher might point out resources to be checked for information. Data used in answering the questions need not necessarily come from actual laboratory work but the students should have to expend energy in gathering the information. Using multiple sources of data will reserve the laboratory work for the information that cannot be gathered elsewhere or to justify the information the students feel may be in error. The teacher's function here is to help the students evaluate the questions and to help them formulate methods for the collection of data.

Once the students realize what data is to be collected, the teacher can assist them in determining the sources to use. It will also be necessary to take a more active role in the formulation of laboratory procedures for safety reasons. When this portion of the instructional process is completed, the student should have recorded the entire procedure for the collection of all data that will be gathered from the laboratory. This should be completed and written in the students' notebooks before they start collecting any data.

The students should have their notebooks with them as they work in the laboratory to permit them to record observations as they work through the procedures they have developed. They should be encouraged to make observations that may be of assistance when they analyze what they have done in the laboratory. Some students will not want to take the time and others will record the observations indiscriminately. Constructive criticism from the teacher should guide the students to include only the unexpected events that may assist in explaining errors. Other criteria can be set by the teacher to improve the general competency of the students in making observations.

The students should take the unorganized data and compile it into usable form. The purpose of this is to demonstrate the value of such organized data in drawing conclusions and answering the given questions. This data may be reduced to tabular form or it may be included in the form of graphs. The students may be asked to do both and to comment on the effectiveness of each mode of presentation with respect to providing information for the questions. The teacher will have the option of working with the students in groups to do this or assigning this task to be done outside of class time. It is recommended that this be done in class until the teacher is confident that the students can successfully handle it on their own.

To assist the students when this material will be reviewed in the future, they should be required to include a list of the mathematical formulas used to treat the collected data. It would also be advisable to show at least one sample calculation using each formula with some of the collected data to assist in correcting any mistakes made in the analysis of the data.

When these steps have been satisfactorily completed, the students will be ready to answer the questions originally posed by them. It is recommended that this be accomplished in either a small group or large group situation. The individual student should not be expected to handle this by himself as he cannot subject his answer to the criticisms that might cause him to think to a greater depth. It will permit wrong conclusions to be rejected by the group or, as a last resort, by the teacher.

The students should be made aware that the investigation processes of science do not stop with the answers to the initial questions. In fact, they should discover that the original answers are only tentative and used only until new and more exact information can be acquired that will provide a more detailed answer to each question. They should spend time analyzing each problem investigated and the procedures used to discover weaknesses and possible sources of error. At this point, the students should be able to make positive statements about their confidence in the information gathered and the conclusions drawn as a result of their work. These feelings should be included in their laboratory books so that they may be inspected and reevaluated at a later time.

Ideally the work done on one problem area will generate questions for the next series of investigations. The students should not be allowed to think that experiments are something that can be completed and then left, but rather that the entire investigative process is continuous. Working in groups of two or more the students can generate questions from their work which could be investigated.

The grading of this type of experimental report will require imagination on the part of the teacher. If the class has been broken into small groups, the teacher may evaluate on a continuous basis by reading the notebook of one member of each group and checking to insure that the others have recorded the same thing. This will require a day-to-day checklist so that the student will always have access to his notebook. Initial work will require that deadlines be established for the completion of the students' work. Time may be saved by checking to insure the work has been done. Constructive criticism should be given to the student(s) as their notebooks are read so that they will understand what mistakes have been made and what steps to take to correct the situation. If the teacher has helped the students to develop criteria against which their work will be judged, the problem of grading is reduced to one of checking to insure that the students are utilizing those criteria.

Independent Projects

There are times when an extended project will develop scientific skills. This will usually come toward the end of the school year. possibly in the last six weeks. At this time a teacher should select or have ready a number of possible areas of research from which the students may choose. Although it is desirable to have an independent project conceived by and outlined by the student, in reality, the research available must be within the area of specialization of the teacher or some other qualified faculty member.

Problems can be constructed that require the students to collect a portion of their data outside of the classroom. This will become increasingly important as the stress on environmental quality causes each community to assess its own individual situation. Problems such as water purity or carbon dioxide level in the air may be researched by small groups after school and on weekends. It is the teacher's responsibility to make sure that the students have a clear and concise set of directions as to what they are to be doing and where they are to be doing it. An attempt should be made to gain the cooperation of industrial and community leaders in making the community "the classroom."

Providing for the ambitious student is probably the greatest challenge the new teacher will face. This particular type of student is interested and willing to do the work, but the teacher is limited by other classes and outside activities. One provision that can be made for these students is to construct "kits" containing the dry chemicals and any glassware not normally found in the laboratory drawer. The kits may be nothing more than shoe boxes. These kits should contain a statement of the problem to be studied as well as a complete set of directions for the collection of the data. A set of open-ended questions may also be included. Depending on the ability of the student, a teacher might just decide to allow the student to choose among various problem statements. The student would be expected to submit a detailed procedure for any laboratory work before being allowed to proceed.

A student should never be allowed to work by himself in the laboratory with any glassware or chemicals. A competent instructor must be present. This severely restricts the times when a student can do extra work. If the physical arrangement of the room permits, the student may be allowed to work while other classes are being conducted. This could most desirably be during a time when the other class is having a laboratory session. It is not advisable for the student to work in the chemical preparation area of the stockroom unless that area is visible to the instructor while the student is working. Other possible times for student work in the laboratory are before and after school depending upon the willingness of the teacher to donate his time for the student. It is not wise for a student to take glassware or chemicals home from school to work on "experiments." If the student obtains the materials from other sources and does the laboratory work without any knowledge on the part of the instructor, the instructor cannot be blamed for any consequences. The teacher should be aware of attempts on the part of some students to obtain glycerin or toluene as well as nitric acid from the chemicals available in the laboratory. The results could be potentially hazardous.

Maintaining an Inventory

The mechanical details of conducting a laboratory are possibly as time consuming as the actual preparation for the academics of each class session. The teacher must be aware of the chemical requirements, glassware requirements, and demonstration equipment requirements. It is also necessary to provide for the replacement of all materials that are either consumed or broken during the academic year. Because of budgetary considerations, requests for monies for expenditures in the next school year must often be made before the end of the first semester of the present academic year. Usually the teacher will find a certain stock of chemicals and glassware that will serve as a core to the program he intends to offer. As suggested earlier, if the teacher will outline the equipment required and the chemicals that will be used in each experiment with the amounts to be used, a summary of these sheets (see examples in the Appendix) will show a projected usage for the coming school year. After checking the need against the present stock, a comprehensive order may be compiled early enough to insure that what is needed will be available when needed. When the teacher desires to deviate from the schedule and select new experiments, the new chemicals used should be added to the list and the old ones deleted. It is advisable to add the capability of conducting at least three new experiments each year until a repertoire of fifty experiments is achieved. These experiments, however, should not require the stocking of exotic or extremely hazardous chemicals. A list of chemicals that should serve as a foundation for any chemistry course may be found in the Appendix.

In selecting experiments the teacher also has the option of determining the size of laboratory glassware to be used. Basically there are three classifications of glassware: macro, semimicro, and micro. Each classification has its unique advantages and disadvantages, and the teacher should consider each if he is given the opportunity to order extensively for the chemistry program.

The macro glassware consists of the larger pieces of equipment, that is, 500 ml Erlenmeyer flasks, 100 ml graduated cylinders, 400 or 600 ml beakers, and the like. If large quantities of reactants are necessary to produce the desired observations by the students, macro glassware is the best to use. This equipment is expensive to maintain as the replacement of larger pieces of equipment as a result of breakage is necessarily more expensive. Sufficient stock of larger glassware should be kept for demonstrations and solution preparation.

The semimicro glassware is somewhat smaller. The student works with 150 to 250 ml beakers, 12 by 150 mm test tubes, 10 ml or 25 ml graduated cylinders. Smaller quantities of chemicals are used so that the advantages of this size of equipment include: (1) smaller breakage cost per student, (2) lower chemical cost per student, and (3) lower initial outlay for the inventory.

The micro glassware is still smaller, utilizing equipment such as 10 and 25 ml beakers, 5 ml or 10 ml graduated cylinders, and 25 ml Erlenmeyer flasks. The quantity of chemicals used and the cost of breakage of this type of glassware is smaller than with the other two types, but it is considerably more difficult for the beginning chemistry student to manipulate the equipment and small volumes used. This size glassware also requires that solid samples be weighed to at least .001 grams, which requires a more sensitive and expensive balance than is usually found in the high school chemistry laboratory.

The semimicro equipment is probably the best choice for stocking the student laboratory work stations. A list of recommended equipment to be kept in each station for student use is included in the Appendix. A list of glassware that should be available in the stockroom for special demonstrations and for preparation of solutions is also included in the Appendix.

Preparing and Dispensing Chemicals

Chemicals for use in any laboratory work will have to be prepared in advance. It is inefficient use of time to require that the students mix their own chemicals unless it is determined that this technique may be gained along with some other insight into solution chemistry. The teacher should prepare sufficient volumes of solutions so that each student may do the required manipulations and have an excess of between 10 and 25 percent. This excess will be used up in inaccurate measurements or by pouring more solution than was needed into the beaker. The student should never be permitted to pour chemicals back into the original bottle. The reasons for this should be explained to the students.

The chemical solutions may be prepared by the teacher or possibly by a student laboratory assistant. In either case bulk storage is encouraged in the stockroom, but both the solids and the solutions should be available in the laboratory. To save time, the student should be provided with two or more bottles containing 100 grams of dry chemicals or 250 ml of solution. The students are usually limited by a fifty-five minute laboratory period, so time should not be used waiting for others to obtain the materials they need. This should be particularly important even if it means that more chemicals are wasted than might otherwise be. All bottles should be plainly labeled so that there is no confusion for the students regarding each bottle's contents. The students should also be encouraged to mark the beakers or test tubes in which the chemicals are placed so that a minimum of mistakes will be made at the benches. When several chemicals are available at the laboratory stations, the students should be aware of any combinations that might produce undesirable reactions. The teacher should point these out and write them on the board before the laboratory work begins.

After the experiment has been completed, excess chemicals should be returned to the stockroom. Enough volume of each solution may be stored on the shelf so that some student could make up the work or use it in an independent project if that solution has a satisfactory shelf life. Dry chemicals should be combined into a separate bottle different from the initial container but plainly labeled. This will guarantee that the original bottle is not contaminated. The excess dry chemicals will usually not be contaminated sufficiently to prevent their future use in most general laboratory work; however, it will also insure available dry chemicals of known purity.

For qualitative work the teacher may want each student to have a reagent rack of small volumes (10-15 ml) of a number of different chemicals available. These reagent racks may be stored at the work station if space and safety permit, or may be stored in another cabinet and obtained by each student when needed. This enables each student to have immediate access to any chemical that is needed. It will require extra work on the part of the teacher to initially stock each of these trays at the beginning of each year and to maintain the volumes in each rack during the year. A laboratory assistant could be used to perform this function. Blocks containing the standard acids and bases that are frequently used may be stored at the work station or in a cabinet elsewhere in the room. It is probably most advisable to have these bottles available in some place other than the work stations to reduce hazards between the laboratory periods. This will also facilitate the replenishing of these chemicals.

Laboratory Design

A laboratory may be designed in many ways to accommodate the laboratory work that will be conducted. The teacher is urged to consult one of the more common methods texts or a good laboratory equipment catalogue such as Kewaunee Technical Furniture Company or the Hamilton Laboratory Furniture Company. In any design, each student should have gas, water, a sink, and electricity available at his work station. Glassware storage and general hardware storage should also be included. The arrangement should be such that it will be easy for groups of two or four students to work on a single procedure if necessary. Consideration should be given to providing two or more carrels complete with gas, water, a sink, electricity and audiovisual equipment so that laboratory work can be completed along with the necessary instruction without the direct supervision or action of the teacher.

Laboratory Assistants

Laboratory assistants can serve to increase the effectiveness of the teacher and the program. Many parts of the programs, such as solution preparation and equipment inventory, can be handled by the laboratory assistant. These assistants are most easily obtained from the ranks of senior level students who are interested in science. This experience would serve to broaden their background and possibly direct them to a career in science or science teaching.

These students should be carefully selected by the teacher in consultation with the department chairman and the student's advisor. The students should have had or be taking the course in which they are to assist. They must have satisfactory marks, and assisting with the class should work no hardship on their present scholastic standing. They should be used for one period a day, preferably their study hall or free time period, and should not be expected to work before and after school as they will not be paid for this work. The incentive for the student to participate is usually based upon his interest as well as the promise of a future recommendation.

The teacher can spend extra time training these student assistants to perform the routine tasks and they will be able to proceed on their own in carrying out assignments. If possible, two assistants should work together, particularly if they are working in the stockroom or preparation area and will not be directly visible to the teacher. Arrangements should be made for them to put in an appearance for safety reasons, at predetermined intervals, such as 10 or 20 minutes, when they are handling chemicals or apparatus that might be hazardous. The teacher must be absolutely convinced that the student is capable and responsible before using him as a laboratory assistant. The assistants can also be valuable to the teacher in the laboratory by looking out for possible hazards, by handling requests for the replacement of broken equipment, or in obtaining chemicals. They may also serve as group leaders for research projects if the teacher is confident of the ability of these students to handle such an activity.

In some school systems the teacher will find that a person is already employed full-time to handle all of the details of ordering chemicals and equipment as well as preparing solutions. These teachers need only familiarize themselves with the procedures followed in that school's system.

Maintaining an Inventory

It is relatively easy to order chemicals and equipment based on projected usage and breakage. When it is assumed that there is no inventory, however, it is more important to base these orders on what is actually needed. An inventory is necessary to serve this function. A separate inventory for equipment and chemicals should be maintained on a master list and on a set of 4 by 6 cards. The inventory from day to day can be maintained on the 4 by 6 cards as the stock comes in and is used. An example of one such card may be found in the Appendix. A master inventory should be conducted once a year for insurance purposes.

A form located near the chemicals should be used to record which chemicals are used. This usage list could be subtracted from inventory cards either weekly or monthly depending upon the individual teacher. Each card should be dated as to when the change was made and initialed by the person making the change. The old forms should then be filed by academic year.

Glassware breakage should be recorded on a form listing the description, size, and number of items that have been broken. It is up to each individual teacher, if no school or departmental policy exists, to decide if the student responsible for the glass breakage will be charged for it. Experience indicates that when semimicro equipment is used, the breakage will vary from $1 to $4 per student per year. The teacher and laboratory assistants should be reminded to fill out breakage forms for any equipment they break so that an accurate inventory may be maintained. The inventory cards for glassware should be updated as the situation demands.

Once the initial inventory is completed, it should be kept up to date. A regular schedule should be followed to record breakage and usage. It should also be changed as soon as new shipments of glassware are received. The numbers for ordering and the company from which they are usually ordered are listed to facilitate reordering. An approximate purchase price may also be included, although these prices are usually not stable from year to year.

Purchasing

The purchasers of additional instructional equipment and materials must consider:

1. What is needed to improve conditions for science instruction?
2. How much money is available?
3. What equipment and materials are available for purchase?

The answer to the first question must be supplied by each teacher after considering his students and the science content/processes that they are to learn from the course of instruction. The school's set of existing equipment, both its quality and quantity, and its relationship to the desired conditions for science instruction should be considered along with possible additions to this set.

The second question, which relates to the size of the budget, is the limiting factor in almost every school's purchasing plans. If local funds are inadequate, supplemental grants can in some cases be obtained from the federal government or from state or private agencies. Check to see if your school administration has any information about funds from these outside agencies.

The types of equipment and materials that are available, along with their prices, can be determined from the advertisements and catalogues of suppliers of scientific equipment. A list of scientific supply houses featuring items of interest to physics teachers is included in the Appendix. Copies of these catalogs can be obtained by a request written on the school's letterhead stationery. Apparatus notes, articles, and advertisements in periodicals such as *The Physics Teacher*, *The Science Teacher*, and *The Journal of Chemical Education* will help the teacher to keep up with new products.

Storage

Some type of storage facilities isolated from student traffic should be provided for equipment and supplies. This will minimize equipment damage caused by unauthorized handling and reduce losses from pilferage.

A storage arrangement that keeps all of the parts of a particular set of apparatus together is desirable. This will save time when setting up the apparatus and can prevent the loss of small parts.

The conditions in the storage environment should be designed to prevent deterioration of the equipment and supplies. Dust covers are desirable for large pieces of apparatus; inexpensive plastic bags from dry cleaners or laundries are well suited for this purpose.

Dust-free storage for lenses, mirrors, and other optical equipment is also desirable; the surfaces of these items should also be protected against scratches. Special consideration should be given to the storage of sensitive items such as cameras, telescopes, and electronic equipment. These items should not be subjected to high temperatures or corrosive environments. Some supplies need special consideration in storage; for example, the shelf life of photographic film can be extended by storing the film in sealed plastic bags inside a refrigerator. Special facilities are needed for the storage of radioactive materials; see Laboratory Safety.

Information Files

A file should be established for all equipment assembly instructions, parts lists, warranties, and operating instructions or manuals. Personal notes on special procedures or techniques that have been discovered should be included; one does not always remember how things were done "last year."

Equipment Maintenance and Repair

1. Upon receipt of new equipment, the warranty conditions should be noted. If a warranty is provided, there is usually a time limit upon it. In some cases, the warranty could be voided by attempts at "unauthorized repair."

2. Preventive maintenance:
    a. Remove all batteries from equipment that will be stored for more than thirty days.
    b. Follow the instructions on cleaning the rollers in Polaroid cameras. If this is done each time a new film pack is inserted, built-up deposits on the rollers and the resulting streaked prints can be prevented.
    c. Lubricate moving parts regularly but sparingly.
    d. Prevent corrosion of unprotected metal surfaces; if necessary, cover these surfaces with a light coating of petroleum jelly or lubricating oil.
    e. Change the oil in a vacuum pump if it becomes contaminated. Observe the manufacturer's instructions on the oil level to be maintained and the normal oil change interval.
    f. Air tracks or tables should be covered to protect them from dust when they are not in use. A vacuum cleaner used as a blower for an air track or table should *not* be used for cleaning purposes; the probability of clogging the fine holes with dust would be too great if the cleaner served both as a blower unit and as a vacuum cleaner.
3. Maintenance contracts with manufacturers for sophisticated equipment, such as analytical balances, should be considered to protect the equipment when the teacher is not an expert technician or where warranties would be voided.
4. Maintain an adequate supply of fuses, bulbs, batteries, and small hardware items. When ordering new equipment, note if any special fuses or batteries are required for this apparatus so that spares can be ordered.
5. A few tools should be available in the classroom for the repair or assembly of equipment. The following items should be included:

adjustable wrench
knife
pliers
both regular and Phillips screwdrivers
soldering iron and rosin core solder
wire cutter and stripper

Many more items such as a hammer, more wrenches, or special pliers could also be useful, but the items listed would suffice for most purposes. A hemostat, a small locking plier available from medical or biological supply houses, is a useful addition that is convenient for work with small parts or as a holder for parts during soldering.

6. Minor Repairs:
   a. Check inoperative electrical equipment for blown fuses or tripped circuit breakers, bad switches or power cords, or loose connections. These simple problems are the most common reasons for electrical equipment failure.
   b. "Pegged Meters," where the needle and movement are jammed at the high end of the scale, can result when the meter is overloaded. Try to purchase meters with overload protection to avoid this problem. However, if overloading does occur, see if there are any screws that can be loosened to give the meter movement more freedom, and then attempt to dislodge the meter movement gently by hand. If it does come loose, straighten the pointer if necessary, reassemble the meter and check its operation with an accurate meter before returning it to service.
   c. Separated mercury columns in thermometers can be reunited by cooling the thermometer until all of the mercury contracts down into the bulb. This can be done by dipping the bulb into a beaker of liquid Freon and then lifting it out and allowing the Freon to evaporate. This may have to be done several times before all of the mercury is drawn into the bulb. If the thermometer is allowed to return to room temperature slowly, the column should be free of gaps. Freon can be obtained in aerosol cans or from your local refrigeration serviceman. Dry ice can be used to cool the thermometers if it is more readily available than Freon, but the Freon permits more gradual temperature changes.

7. Major Repairs: Consult the manufacturer of the apparatus. Most items will have to be returned to the manufacturer or to an authorized service agency for major repair. Compare the repair cost plus shipping charges against the purchase price of a replacement before deciding upon a major repair job.

Laboratory Safety

The teacher must be aware of the types of accidents that can happen in the laboratory and should take all possible precautions to insure the safety of the students. In spite of the best precautions and continuous monitoring there will be times when accidents occur. The teacher must be ready to act immediately and correctly in any situation to prevent extensive harm to the student. The student must also be ready to act to help minimize damage to others as well as to himself.

Teachers have recently been held legally responsible for accidents that occurred in their laboratories. It is not the author's intention to frighten, but merely to inform the teacher of the possibilities that may occur in conjunction with fulfilling the duties of a chemistry or physics teacher. The teacher should obtain a copy of *Science Teaching and the Law*, by Brown and Brown from the National Science Teachers Association. This booklet examines most of the accident possibilities and gives many recommendations that should be followed. Awareness, constant supervision, emphasis on safety, and appropriate and immediate action will reduce the possibilities of legal action as well as bodily harm.

One of the best ways to prevent lawsuits is to conduct the laboratory in such a way as to minimize the possibilities of accidents. The teacher should also obtain a copy of *Handbook of Laboratory Safety*, by Norman V. Steere (ed.), published by the Chemical Rubber Company. This publication provides detailed instructions for handling all types of chemical hazards that may be present in the laboratory. Other publications are listed throughout this chapter.

Articles on safety and laboratory problems appear frequently in the *Journal of Chemical Education* and *The Physics Teacher*. The teacher should search out these articles on a monthly basis and copy the suggestions for future reference. Other magazines may also publish articles on this topic, and the teacher should try to read as many of these as possible.

A thorough knowledge of first-aid procedures enables the teacher to render the appropriate treatment. Ideally a first-aid course should have been taken in college, or the teacher should enroll in such a course at the earliest opportunity.

The following list of rules for the teacher is presented with the intention that the teacher become aware of his responsibilities and modify his conduct in the laboratory accordingly. This list contains the summation of years of experience by many chemistry and physics teachers. Most of these rules are common sense, but their implications have value and should therefore be considered.

Rules for the Teacher

1. Record all accidents.
2. Never leave the students unattended in the laboratory while they are performing experiments.
3. Provide the students with safety rules and remind them frequently of what is expected.
4. Know the experiment.
5. Know your students.
6. Make sure that the students dispose of used chemicals in the most appropriate manner.
7. Instill an atmosphere of safety.
8. Inform and review the procedures for fires, accidents, and other emergencies on a regular basis.
9. Provide a safe place to work.
10. Never use students to do your work.
11. Never allow unauthorized students in the stock room.
12. Inform all students of the location of all safety equipment.
13. Teach good work habits.

1. RECORD ALL ACCIDENTS.

The teacher should keep records of all accidents that occur in the laboratory. Even minor burns and cuts should be recorded so that appropriate action can be taken immediately and records will be available if side effects become apparent at a later date. A 3 by 5 card such as Figure 4-4 can be used for the initial record. A more detailed listing should then be prepared with copies being sent to the department head and to the principal if required by the school.

If the teacher finds that he is frequently experiencing the same type of accident, he should examine the laboratory as well as the experiments to see if that particular hazard can be eliminated.

The teacher may be tempted to dismiss minor cuts and burns as insignificant. This will depend on the student. Occasionally there will be students with health or physical problems that could reduce their ability to act safely in the laboratory. The teacher may not be aware of these disabilities. A written record will provide the teacher with the details that may be needed at a future time should the student or his parents decide to sue for damages.

```
Date:                                    Time:

Name:
Nature of injury:

Treatment given:

Follow-up:

Cause of accident:
```

Figure 4-4

2. NEVER LEAVE THE STUDENTS UNATTENDED IN THE LABORATORY WHILE THEY ARE PERFORMING EXPERIMENTS.

This applies 100 percent of the time. The teacher must be aware of what is happening during the course of an experiment. It is enough of a responsibility to supervise twenty-four students handling chemicals without assuming that they are capable of working without supervision. Cases have been reported in which students were injured by the careless use of chemicals while teachers were out of the room. These same teachers were later held liable in court for damages. The experienced eye can see potential hazards and correct them before damage is done. Usually the laboratory period will only be fifty-five minutes in length. The teacher should, therefore, be able to wait until a break between classes if there is a need to leave.

Caution should be used if trips to the preparation or stock room become necessary. Occasionally a chemical will be exhausted or a piece of glass will need to be replaced before the end of the hour. The teacher should check at the end of each class period to be certain that there is a sufficient quantity of chemicals and equipment for the next group of students. It is also advisable to keep a few pieces of extra glassware in the laboratory to replace that which may be broken in the laboratory. Students should not be permitted to go into the stockroom or preparation room to obtain glassware or chemicals. The student unknowingly may get the wrong chemical and use it, or he could steal some potentially harmful chemical without being detected. If the school provides laboratory assistants for the teacher, they should be used for this purpose as well as in the capacity of additional supervisors to prevent accidents.

3. PROVIDE THE STUDENTS WITH SAFETY RULES AND REMIND THEM FREQUENTLY OF WHAT IS EXPECTED.

A list of general safety rules will be considered later in this chapter. The teacher is urged to examine this list in detail and add any other rules that will fit the individual situation. The important objective is that safety consciousness be developed from the first day. The rules should be duplicated for inclusion in the students' notes and copies should then be posted around the room. Reminders may then come in the form of films, posters, quizzes, or verbal instructions that accompany the discussion of an experiment and by other means as they become appropriate for the situation.

4. KNOW THE EXPERIMENT.

Advanced lesson preparation for each experiment should include a review of the lesson content and the safety precautions that are to be observed. If the teacher has prepared a chemical-use list for each experiment, this should be used as a starting point. The teacher should then examine the health hazards of each of the chemicals being used and what to do if these chemicals come in contact with the student's skin or are ingested. The *Handbook of Laboratory Safety* by Steere contains tables of the commonly used chemicals and their relative hazards and antidotes. This list should be available to the teacher in a handy location in the laboratory. If the teacher has familiarized himself with the hazards before class, action may be taken at once if something should happen. This will also insure that any unusual antidotes will be on hand in case of an accident.

The teacher should also prepare a list of the possible mistakes that the students might make during the experiment. It should be remembered that students do not have all the skills to handle chemicals. The high school chemistry course should be teaching these skills as well as providing valuable physical experiences to which the students can relate the content of the course.

The students should be made aware of any new or unusual safety precautions to be observed before the experiment is begun. The teacher should have a list of these safety rules prepared and included with the previously described lists. The proper emphasis may be obtained by reading the new rules to the students. A better effect can probably be obtained by handing out copies of the rules and then reading them to the students.

The teacher may wish to develop procedures for collecting data in the laboratory. If a standard experiment is followed, these procedures will be outlined in great detail. If the teacher desires to pursue an inquiry approach, it will then be necessary for him to have a guideline prepared in advance so that the inquiry does not proceed unless the procedure has been sufficiently developed to insure safety during the collection of data.

5. KNOW YOUR STUDENTS.

At the earliest possible opportunity, the teacher should obtain a check on the medical history of each student. If the student has a physical or medical handicap, it is better to know this in advance so that extra precautions may be taken. Students should inform the teacher if they are taking medication under a doctor's orders. If a student is subject to seizures on a regular basis, the student's actual laboratory work should probably be reduced to the role of an observer with another student or group of students. Most counselors will co-operate in obtaining these medical records once the safety aspect is explained to them. Do not expect the counselors to voluntarily provide this information. Students may also be asked to report such handicaps before they begin work. The teacher should strive to prevent the student from being embarrassed, and all information received should be treated in a confidential manner.

6. MAKE SURE THE STUDENTS DISPOSE OF USED CHEMICALS IN THE MOST APPROPRIATE MANNER.

Disposal of the waste products from the laboratory has not received the attention that it should. Chemicals are often flushed down the drain without any thought to the possible hazards that might be created.

Solid materials that have been collected in glass crocks are either included in the trash disposal from the school or thrown out on the ground away from the building without respect to possible health hazards. The teacher will have solid and liquid wastes to remove from the laboratory. It would be unusual if any noxious fumes were generated in sufficient quantities that their discharge into the air would create a hazard. However, the teacher should take the possibility of noxious fumes into account and use hoods if necessary. The teacher should be aware of the location of the discharge of the hood ventilation system and where the fumes might travel from there. A simple test with hydrogen sulfide should indicate what areas of the building might be affected by the discharge of gases during an experiment.

The temptation that leads to a large number of laboratory fires is allowing the solid residue to accumulate until there is sufficient bulk to make disposal worthwhile. Although many solid chemicals will not react with each other in the solid form, a student will occasionally throw in a slurry of chemicals that provide enough water for an unpredictable exothermic reaction that could produce any number of products. The teacher should make the student aware of this disposal problem and insist that all reaction products from a given experiment be collected at one point and then disposed in individual containers. If the teacher is aware of all of the reaction products, (and a detailed preliminary analysis of the experiment should provide this information) their relative reactivities along with their physical properties and other pertinent data should indicate the appropriate method of disposal to use. Excess water may be removed by filtration and/or evaporation and the residue may then be transferred to small disposal containers. Glass jars that can be transferred to an approved disposal area will serve for this purpose.

Burial is usually the most satisfactory method of removing waste products from the environment. Open dumping could lead to contamination of the ground water by passing into the water supply of the immediate vicinity of the dumping grounds if the solids are not stored in leak-proof containers. For the volume of waste that the average high school laboratory will produce it might be impractical to maintain a separate chemical dump if other dumps are available. If a local company is conducting chemical research or using large quantities of chemicals in its industrial process, it is possible that arrangements could be made to use that company's available disposal facilities. If not, the school should provide a concrete burial pit for these chemicals. The local health department should be consulted as to the proper location and construction of this facility so that it will comply with the state and local ordinances.

Liquid waste disposal presents other problems. If the volume is small, the logical tendency would be to flush it down the drain with large volumes of water. The advisability of this depends upon the relative toxicity of the ions being removed from the laboratory in this fashion. Again, advanced planning by the teacher will provide him with a guide line of the best procedure to use at this point. It should be remembered that all disposal of liquid wastes should comply with federal, state, and local laws. Flammable liquids such as alcohol or ether should not be flushed down the drain because of the possible buildup of fumes that could set off an explosion. Organic liquids could either be buried in appropriate containers or burned in an open area. The burning process used should insure that complete combustion takes place. The teacher should rely on the laws, the toxicity, and the quantity of liquid waste to be removed from the laboratory in selecting the appropriate method of disposal.

7. PROVIDE AS SAFE AN ENVIRONMENT AS POSSIBLE.

The teacher's attitude and habits will determine what his students learn and how the students will work in the laboratory. If safety is a genuine concern, they will tend to be safety conscious. The teacher must also guard against instilling fear in the students. This will come with experience and will be manifested in how the teacher works with the chemicals and the interest shown in laboratory work. The teacher must be willing to let the students use procedures that they have developed to collect the data necessary to evaluate the hypothesis in question. To do this will require an understanding of the problem being considered and a constant supervision of the students to insure that they do not proceed in a manner that might lead to an accident.

8. INFORM AND REVIEW THE PROCEDURES FOR FIRES, ACCIDENTS, AND OTHER EMERGENCIES ON A REGULAR BASIS.

Each school has its own regulations for reporting fires and evacuating the building should a fire occur. The teacher should familiarize the students with these rules early in the year so that they will be prepared for a fire drill. Accidents in the laboratory and the procedures to be followed should receive emphasis when the safety equipment is being discussed.

Of particular concern to the teacher is the establishment of procedures to be followed in the event of a fire elsewhere in the building or a fire drill during a laboratory period. The chemistry teacher cannot have the students leave their stations unless all possible hazards have been eliminated; that is, Bunsen burners must be off, reactions stopped, and any hazardous chemicals returned to a safe place. If possible, the teacher should ask to be informed in advance if a fire drill will be held; if the principal will not provide this information the teacher could provide him with a schedule of when his classes will be working in the laboratory. The principal should also be impressed with the possible hazards involved in a hasty shutting down of some experiments. The teacher should not allow himself to be hurried in getting his students out of the building if the laboratory will become a hazard once it has been left.

Shut-down procedures should also be stressed for those times when the teacher's attention is focused on one or more students as the result of an accident. If the teacher will be occupied in this manner for more than five minutes, the students should be instructed to stop work and return the laboratory to the pre-experiment conditions. If the teacher must leave the room with a student, the class should stop work immediately. The teacher will have to take the chance that the cleanup operation will proceed without incident. If time permits, a teacher in an adjoining classroom should be informed of the situation and asked to supervise the cleanup if possible. In all accidents the safety of the student is the primary concern. This includes the student who may need medical assistance as well as those remaining in class.

9. PROVIDE A SAFE PLACE TO WORK.

The overall responsibility for safety in the laboratory rests with the teacher. It is his responsibility to provide a safe place to work. This may be accomplished by providing the students with access to only those chemicals needed to complete the experiment being conducted. It also will help to remove chemicals from previous experiments to the stockroom for reshelving or disposal by an approved method. Cleanliness should be stressed in each class, and the laboratory bench itself should be cleaned by the student working within that area. If student help is available, one of their responsibilities should be to wash all counter spaces and to return excess chemicals to the stockroom. All acids and bases as well as other chemicals should have their place in the laboratory, and the students should be informed that chemicals are not to be removed outside of these areas. Bulk quantities of chemicals should be confined to an area, and the student should draw the quantity needed for the day's work and return to his individual station. The student should not be allowed to have the bottle quantities at his station because this will provide a greater quantity than needed. The student will often fail to return the bottle to the main dispensing area and other students will waste time trying to find the chemical. Since the laboratory periods are relatively short, any time lost decreases the benefits the students will gain from that day's activities.

10. NEVER USE STUDENTS TO DO YOUR WORK.

It was mentioned earlier that only student assistants should be allowed to obtain glassware and chemicals from the stockroom. Even these students should be used with the greatest of caution. They must be trained and capable of performing the required tasks and responsible enough not to venture into areas where they are not sure of themselves. There will be times when a willing student might be used to run an errand such as being sent to purchase a chemical from a drugstore. The teacher is discouraged from using students in this manner and urged to check the school policies on such actions. Teachers have been held liable for accidents occurring to students while they were performing errands for the teacher. It is far easier for the teacher to do it himself than to risk what might happen. Most schools have other channels through which supplies may be obtained.

The teacher will also have students with a genuine interest in the subject who desire to do extra work around the laboratory. Hopefully these students can either be used as laboratory assistants or a course of independent study can be worked out for them. The teacher should not allow these students to help with preparations unless such preparations are under the teacher's direct supervision. In no case should they be allowed to work by themselves.

11. NEVER ALLOW STUDENTS IN THE STOCKROOM.

As mentioned earlier, it is not a good practice to allow the students in the stockroom. There are many safety reasons for this as well as liability reasons. Students who show an unusual interest in drugs or chemicals should be watched to insure that they do not remove chemicals from the laboratory or from the stockroom. If the stockroom contains a large number of organic compounds. the teacher should keep a frequent and accurate inventory to insure that students are not stealing chemicals that could be used to produce such drugs as LSD. Student workers should be allowed in the stockroom only if they have proven to be trustworthy and responsible. Even then they should not be allowed to work alone or unsupervised at frequent intervals because of the possible health and safety hazards.

12. INFORM ALL STUDENTS OF THE LOCATION OF ALL SAFETY EQUIPMENT

The teacher will not always be in the best location in the room to assist with the immediate solution required for a given situation. However, a student will be near the appropriate safety equipment and, therefore, all should know the location of the equipment, and when and how it is to be used. The fire extinguishers, fire blanket, eye wash, safety showers, and first-aid kit locations should be familiar to each student in every class. The teacher is still responsible for the actions of the students and should take charge of the situation as quickly as possible. Time spent in the instruction of the use of these safety devices can prevent serious damage to the students in the event of an accident.

13. TEACH GOOD WORK HABITS.

Many accidents will be prevented if the student is taught to work carefully and safely. The teacher can do much to create the proper atmosphere by being safe and conscientious in the laboratory. Orderliness, cleanliness, precision, knowledge of what is to be accomplished, what procedures to use, and knowledge of how to handle equipment and chemicals and the safety rules are factors that will enable the student to work effectively and safely in the laboratory. Possibly the most important attribute to instill in the student is that of respectful caution. The student must be made aware of the existing hazards and that these hazards may be overcome by the use of proper procedures.

The following list of rules for the student is to be applied when the student is working in the laboratory. Most are a matter of common sense but as the student becomes familiar with laboratory work he will have a tendency to forget some of the rules that provide for his safety. Space has been left at the end of the list to enable the teacher to add some additional rules that may apply to an individual situation. The teacher is encouraged to make a copy of these rules available to the students. Reminders in the form of quizzes and posters should instill in the students the necessity for constant vigil.

Rules for the Students in the Chemistry Laboratory

1. Always wear safety glasses when handling chemicals. Wear aprons or gloves as the instructor advises.
2. Keep your appearance as neat as possible. Wear snug fitting clothing and keep long hair tied back behind the head.
3. Flush all chemicals off the skin and clothing immediately with plenty of water.
4. Do not point test tubes at your neighbors while heating liquids of any kind.
5. Do not hold your face over an open container. All fumes and odors should be observed cautiously.
6. Use only the prescribed quantities of chemicals. Never attempt unauthorized experiments.
7. Insert glass tubing and thermometers into rubber stoppers with great caution. Follow the directions of the instructor.
8. Dilute acids by adding the concentrated acid to distilled water.
9. Never return unused chemicals to the original containers.
10. Dispose of extra and waste chemicals as directed.
11. Do not taste any chemicals in the laboratory.
12. Be cautious with heated glass. It may not be as cool as it looks.
13. Use the hood for experiments that produce noxious gases.
14. Restrict conversations to the experiment. Socializing may be conducted after class.
15. Know the location of all safety equipment.
16. No "horse play" in the laboratory at any time.
17. Inform the instructor of all accidents no matter how minor.
18. Know what you are doing. If you don't, stop and consult the instructor.
19. Do not bring any food into the laboratory.

1. SAFETY GLASSES.

Most states now require that safety glasses be worn by students in science laboratories and shop classes. The teacher should contact the state science association if he has any doubt. The type of safety glasses that must be worn is also regulated in some areas. Ideally full face shields should be worn, but the teacher and/or school will usually settle for only protection for the eyes. Regular glasses are not sufficient unless they are constructed of safety glass that will not shatter on impact. Even then they should have side shields that will protect against liquids entering the eyes from the side. It will often be necessary to remind the students that they are expected to wear the glasses as long as anyone is working in the laboratory. Stiff penalties should be invoked if the student continues to avoid wearing the glasses.

Although aprons should be worn by the students, they are not required under state laws at this time. Aprons should be provided in the laboratory at no expense to the student or at only a small rental charge. The aprons will protect the students' clothing against spills of liquids and will enable most splatters to be washed off immediately without further injury to the student or his clothing. The aprons chosen should be acid resistant and of such quality to insure they will last for at least one year under normal conditions. Other protective garments should be worn as warranted by conditions. No experiment should be assigned to the students unless such safety equipment is readily available.

2. PERSONAL APPEARANCE

Personal appearance should be emphasized at the beginning of the year, and it will probably not be necessary to mention it again if the teacher provides a rationale for the special dress rules in the laboratory. Snug fitting clothing is encouraged in the laboratory to keep sleeves and shirttails out of the burners and beakers. Hair should be tied behind the head if it is long enough to swing in front of the face to prevent hair from being burned as a result of contact with a burner. The wearing of sweaters in the laboratory should be discouraged because of the fire hazard with loose yarn or fuzz, and because they are readily attacked by acid splatters.

3. FLUSH CHEMICALS WITH PLENTY OF WATER.

The best first aid for most accidents that may occur in the high school chemistry laboratory as a result of harmful chemicals contacting the skin is to immediately flush the skin with plenty of water. The amount of water used will depend on the extent of the contact. If the spill or splatter of the chemical is slight, then tap water will probably be sufficient. If the spill is large, then a shower or other device that can provide a large volume of water in a short time should be used. The student should be encouraged to wash not only the skin area that has been affected, but the surrounding clothing as well. Only after this preliminary step has been taken should the student seek first aid from the teacher.

4. DO NOT POINT TEST TUBES AT YOUR NEIGHBORS.

When the students are first learning to handle test tubes and Bunsen burners, some may tend to ignore the direction in which the test tubes are pointing. The teacher may wish to demonstrate how far water from a test tube will travel if it is blown from a test tube that is heated in the wrong way.

5. DO NOT HOLD YOUR FACE OVER AN OPEN CONTAINER.

The student should be instructed in the proper way to observe odors. Not only will some odors penetrate the sinuses causing discomfort, but some odors may produce nausea. If the student is too close to a beaker that is being heated, vapors may rise that may injure eyes or lung tissue. The students should be instructed to take only small sniffs and not to bring the liquid close to the nose on the initial attempt to determine the odor.

6. USE ONLY PRESCRIBED QUANTITIES OF CHEMICALS.

Waste of chemicals results when students have to transfer chemicals to a beaker and then accurately weigh the required quantity on balances. Time and chemicals can be saved if more jars of chemicals are available. The student should also be given experience in estimating volumes and masses. Any excess chemicals should be utilized by another student if possible.

Some students will use more chemicals than necessary. They may want to see what larger quantities will do, and sometimes they are just careless. The teacher should strive to have the students use the exact amounts of chemicals to eliminate hazards that may result in either case.

7. NEVER ATTEMPT UNAUTHORIZED EXPERIMENTS.

The student should be made aware of the potential hazards involved if he attempts procedures without first having them verified as safe. If the student has a genuine desire to observe a chemical phenomenon, this should be handled as a special project under appropriate supervision to insure it will be safely conducted.

8. INSERT GLASS TUBING IN RUBBER STOPPERS WITH CAUTION.

Most laboratory manuals will have specific directions for inserting glass tubing and thermometers into rubber stoppers. Most procedures suggest that the glass be coated with glycerine and wrapped in a towel and pushed through the stopper with a steady force. Quick pressures are to be avoided as they will enhance glass breakage that could result in the glass piercing the hand. Thermometers should not be inserted into rubber stoppers unless it is absolutely necessary, since they will often become stuck if left in the stopper for too long, and they frequently break when they are removed. If possible, the thermometer should be inserted in a cork if the experiment will not be affected. The teacher may decide that time may be saved and accidents avoided if he or an assistant inserts the glass tubing in the stoppers.

9. DILUTE ACIDS BY ADDING THE CONCENTRATED ACID TO DISTILLED WATER.

Students will violate this rule as frequently as they will any other. Students will forget which liquid is to be added to the other and will usually forge ahead with only a 50 per cent probability of being right. This rule should be reviewed every time the student is required to dilute an acid. Posters may be displayed on the wall demonstrating the appropriate dilution technique to provide the student with a visual reminder.

10. NEVER RETURN UNUSED CHEMICALS TO THE ORIGINAL CONTAINERS.

The student should be frequently reminded that any excess chemical is not to be returned to the reagent bottle. If the student has removed more chemical than was required, the excess may be saved for future use in a bottle that is clearly marked for that purpose. The teacher should use chemicals obtained in this manner with care as these chemicals will not be as pure as those obtained directly from the bottle. Disposal of excess chemicals should be in accord with procedures developed to handle waste chemicals.

11. DISPOSE OF EXTRA AND WASTE CHEMICALS AS DIRECTED.

A detailed procedure should be followed and posted where the students can readily see it. If the teacher wants to retain some wastes separate from others, this fact should be made known to the student, and the students should be monitored to prevent unwanted wastes from getting in each container.

12. DO NOT TASTE ANY CHEMICALS IN THE LABORATORY.

Some of the elementary chemistry experiments require the tasting of some substances. The safest approach is to avoid such experiments and to indicate to the students that if they want to taste things they should enroll in a gourmet course. The students should also be discouraged from using laboratory glassware as drinking water containers.

13. BE CAUTIOUS WITH HEATED GLASS.

Several students will be burned when they begin working with glass because they will fail to realize the length of time required for the glass to cool and that the glass itself will not look hot when it is. They should also be instructed to allow the glass to cool slowly in the air and not to run cold water over the glass in an attempt to speed up the cooling process.

14. USE THE HOOD FOR EXPERIMENTS THAT PRODUCE NOXIOUS GASES.

Activities that are conducted in the hood will include special experiments and the mixing of flammable chemicals. Mixing acids with water in a hood equipped with a safety shield will also reduce possible odor and splattering damage. The high school laboratory will usually not have sufficient hood space to allow the entire class to conduct work at the same time so that the hood will have restricted usage for special experiments and demonstrations. If noxious fumes might be emitted during the experiment, another experiment should be selected rather than assuming that the experiment will be uneventful.

15. RESTRICT CONVERSATIONS TO THE EXPERIMENT.

After the experiment is underway, the students may find that there is time for conversation. This is to be discouraged unless the conversation is directly connected with the work being done. If the student's attention is distracted, he may lose control of the experiment or fail to note some observations that might have been valuable. This is not intended to discourage small group experiments. Such groups can generate many questions and answers, and make more observations than a single student would.

16. KNOW THE LOCATION OF ALL SAFETY EQUIPMENT.

The student should be informed of where all equipment is located and how it is to be used in most situations. It should be impressed upon the students that prompt action on their parts will minimize most accidents. They should also be encouraged to render assistance to a fellow classmate should the need occur. Quizzes and drills may be used to stress the importance of this knowledge.

17. INFORMING THE INSTRUCTOR OF ACCIDENTS.

This rule should be used in accordance with the teacher's own realization of the need to record any and all accidents as they occur. Since the students will attach as much importance to this as the teacher, the teacher should indicate the concern that will motivate the students to respond accordingly.

18. KNOW WHAT YOU ARE DOING.

The teacher should encourage the student not to try to do anything unless the student is sure of what he is doing. Some students will be afraid that they may appear ignorant and will continue their activity even when they are in doubt. The teacher must seek out these students and encourage them to ask questions when they are uncertain. The teacher should also discourage some students from attempting unauthorized experiments.

19. NO "HORSEPLAY" IN THE LABORATORY AT ANY TIME.

Students should be discouraged from any types of activity in the laboratory that may cause injury or damage. This includes both the laboratory and lecture portions of the course. When the student is in the laboratory-classroom, any unwarranted actions could cause accidents. The chemistry and physics students will usually follow the rules if they are impressed with the possible hazards. If another class is to use the laboratory, the teacher of that group should be informed of the special hazards to the students should they investigate the contents of the laboratory without directions. For these and other reasons, dangerous chemicals and delicate equipment should be stored in a stockroom or in a locked cabinet except when in use. Glass front cabinets with locks may be used to display the equipment without the danger of the equipment being mishandled by the students.

20. DO NOT BRING ANY FOOD INTO THE LABORATORY.

This matter should need no emphasizing, but experience indicates that students will try to bring snack foods into the laboratory if they are not prohibited from doing so. The teacher must inform the students of the potential hazards that may be encountered from this practice. The students should also be required to wash their hands with soap and water upon the completion of all laboratory experiments in which chemicals were handled.

If laboratory assistants are used they should also be warned. A strict rule should be issued on this subject and rigidly enforced. It is also up to the teacher to set the example by refraining from eating in the laboratory when students may observe him.

Characteristics of a Safe Laboratory

1. Adequate student work space.
2. Cleanliness and neatness.
3. Adequate lighting.
4. Adequate safety equipment.
5. Appropriate student/teacher ratio.
6. Adequate ventilation.
7. Waste disposal containers.

Safety in the laboratory will be enhanced by the physical layout of the room itself. Conditions within the room can promote fewer accidents such as spacing between benches and the location of chemicals. Some laboratories have been forced to accommodate more students than they were designed to handle, which may lead to potential hazards for the teacher as well as the students.

Ideally, the number of students working in a high school chemistry laboratory should be restricted to twenty-four, since experience indicates that this is the maximum number that the teacher can supervise at one time and still provide adequate instruction and supervision. If assistants are available, this number may be expanded slightly. In no case should more students be crowded into a laboratory than the number for which it was designed.

The teacher will not usually have the option of deciding what types of benches, i.e., stand-up or sit-down, will be provided in the laboratory. He must do the best he can with the available equipment. If a new laboratory is being constructed or a new laboratory arrangement is being made, careful consideration should be given to requesting benches and adequate spacing as well as physical layout to provide maximum safety.

Sit-down benches have the advantage that they may double as the student's class desk as well as the laboratory work station. Precautions must, however, be taken to keep the student from spilling chemicals in his lap during experiments. Accidents will happen while the student is in the process of sitting down or standing up since his mind is usually on performing the action and not necessarily on what he is holding in his hands. This presents dangers to the students working at adjacent benches, who cannot anticipate his accident nor move out of the way quickly enough to avoid being affected by spills. The teacher can counter these problems by requiring that students wear laboratory aprons and safety glasses at all times. An analysis of the experiments should be conducted and a selection made that will minimize student movement around the laboratory. Chemicals that are to be used may be provided at the student's work station, or all needed materials may be acquired at the beginning of the hour and wastes disposed of at the end of the hour.

Stand-up benches eliminate a number of the accidents that can occur at a sit-down bench. Liquid chemicals will not form a pool in the student's lap, and the time needed for reaching water to flush the chemicals is reduced considerably. If the student is wearing an apron, the liquid will run off onto the floor and the student will usually not even get his clothes wet. If an extended laboratory period is available, there will be a temptation to provide stools for the students. If this is done, the probability of an accident is increased.

In the design of the laboratory, a number of accidents that are caused by student movements can be eliminated by providing a work space of approximately four feet per student and allowing a space of forty-two to forty-eight inches between the desks to allow students to pass. Providing water at the student's desk will also reduce student movement and hence potential accident situations.

Classroom ventilation should be a major concern of the teacher because if it is inadequate, the selection of experiments for class use will be limited. All vapors that may be generated are potentially hazardous if the concentrations are allowed to increase without limit. If a new building is being planned, the teacher should suggest that a manually operated exhaust fan be installed in addition to the hoods and regular ventilation equipment. Some architects will say that this equipment is not necessary, but it is and worth the added cost. If fumes become a problem in an older building, the teacher is faced with the decision of opening the windows for ventilation regardless of the weather or persuading the school administration of the need for installing an exhaust system. If both solutions are impractical, care should be exercised in the selection of experiments. A manually controlled exhaust system is also desirable for the stockroom and preparation room.

Accidents may also be prevented by the appropriate use of the shut-off valves for the gas, electrical, and water supplies for the student benches. The teacher should be aware of the location of these valves and use them when the laboratory is not in use to eliminate accidents caused by unauthorized use of the laboratory. The gas should be turned off each day and all valves for gas, electricity, and water should be shut down when school is not in session during vacation periods.

The initial layout of the laboratory should consider the location of safety equipment and student traffic patterns. The traditional pattern of construction consists of a rectangular room with laboratory benches in the center and bench space or cabinet storage arranged around the perimeter of the room. Chemicals should be dispensed at selected points around the room to eliminate the concentration of students in any one area of the room. This is particularly important if the students would otherwise concentrate close to a work area of another student. Increasing the number of chemical dispensing stations also reduces the time that the student must wait for chemicals.

Safety equipment such as fire extinguishers should be located to provide access from a number of different directions. The normal laboratory should have at least two five-pound dry chemical fire extinguishers located at the opposite ends of the room. These may be supplemented by having aerosal fire extinguisher cans on the laboratory work benches and a bucket of sand in a corner of the room. The use of these extinguishers should be demonstrated at the beginning of the school year to each class. Even though regular inspection of fire extinguishers is required by law, the teacher should inspect every extinguisher at the beginning of the school year and periodically thereafter to insure that each extinguisher is in operating condition and contains a sufficient charge to handle a small fire.

The new teacher or the teacher who is moving into a new laboratory should make an inspection of the facility and inventory of available safety equipment. Each laboratory should contain the following items:

1. Fire extinguishers
2. Fire blankets
3. First-aid kits
4. Eye wash
5. Safety shower
6. Hoods

The first four items are an absolute necessity and the last two should be included if at all possible. In an older laboratory it would be best to use the small individual eye wash bottles that are available. If a safety shower is not available and one cannot be provided, the teacher could improvise by keeping a bucket of water available in a prominent place, which should be checked regularly to make certain that it will be full of water when it is needed. If the school will not install hoods, a portable unit may be used. If this is not available, experiments will have to be used that will not produce toxic or noxious fumes.

The first-aid kit should be available in the laboratory with access by the students as well as the teacher. The teacher should purchase one of the commercially available kits instead of accumulating its contents. In evaluating the purchase of a kit the teacher should examine its contents. Each kit should include the following materials in some satisfactory quantity.

1. Gauze Bandages (2 in. x 126 in.)
2. Gauze Bandages (3 in. x 126 in.)
3. Gauze Bandages (1 in. x 126 in.)
4. Gauze (8 ply, 4½ in. x 36 in.)
5. Cotton (1 oz.)
6. Air Vent Tape—Clear (1 in. x 150 in.)
7. Gauze Pads (2 in. x 2 in.)
8. Gauze Pads (3 in. x 3 in.)
9. Gauze Pads (4 in. x 4 in.)
10. Adhesive Bandages (1 in.)
11. Amoply Ammonia Inhalants (4 cc)
12. Tongue Depressors
13. Eye Pads
14. Swabs
15. Scissors
16. First-aid Cream (1.5 oz.)
17. Tweezers
18. Eye Lotion (3.8 fl. oz.)
19. Adhesive Tape (2 in. x 180 in.)
20. Triangular Bandage
21. Petroleum Jelly

As stated earlier in this chapter, it is necessary to evaluate the quantity and type of chemical to determine the most satisfactory method of disposal of expended chemicals. If solutions are to be collected, a separate bottle should be provided for each different solution or reaction mixture that is collected. Dry chemicals should be collected separately and buried in approved burial sites if they cannot be reused. Mixing of dry chemicals from several experiments can be hazardous if some student accidentally pours a quantity of water into the collecting vessel. The dry chemicals should not be allowed to accumulate from week to week but should be disposed of after each experiment to prevent confusion and possible undesirable consequences. Broken glassware should be collected in a separate container; it should never be placed in waste baskets or included with the dry chemicals. The custodian can usually dispose of the glassware when the container becomes full. He is the one who benefits by your actions to separate the glassware from the paper in the wastebasket.

Characteristics of a Safe Preparation Area

1. Adequate work space and utilities.
2. Sufficient ventilation.
3. Clean and neat work areas.
4. Every chemical and solution labeled.

The preparation area may not be available in all schools and in some it may be combined with the stockroom. Each of these areas possesses certain hazards that the teacher must be aware of in advance. This is particularly important if student help is used in the preparation of chemical solutions. A concentration of chemicals and the nature of solution preparation require that the teacher or assistant observe the safety regulations with respect to the use of safety glasses, aprons, and rubber gloves. Safety shields and a hood for mixing acids should be available and used with regularity. A fire extinguisher should be available in the immediate vicinity in case of fire. The teacher should not work alone or allow any student to work alone in this area. If the teacher is working alone, a colleague should be aware of this fact and should check on the teacher at regular intervals. The hood may be adequate to remove unwanted fumes; if it is not, a window should be opened or some other exhaust system employed to reduce this hazard.

Characteristics of a Safe Chemical Stockroom

1. Separation of organic and inorganic chemicals.
2. Separation of liquids and solids with liquids on the bottom shelves.
3. Quantities of less than two liters or five lbs. if the stockroom is heated or close to classrooms.
4. No heat.
5. Good ventilation.

Storage facilities must be provided for both glassware and chemicals, and it is a common practice to include both in the same general area. Although it may be convenient to store them together, each has special storage problems that must be recognized.

The glassware may be stored in any space but shelving and/or cabinets designed for glass storage should be utilized. The shelves should slope slightly downward from front to back, and should have a small lip on the front edge to prevent breakage of bottles when they are accidentally dislodged. A rough coating of paint on the shelves will also decrease breakage.

The storage of chemicals presents a different problem. Separation should be made between organic and inorganic chemicals as well as between liquid and solid chemicals. Chemicals should be stored on shelving similar to that described for the glassware. Solids may be stored above floor level whereas liquids should be stored on the bottom shelf. Quantities larger than five pounds of solid chemicals or more than three liters of liquid should not be stored in the storage room closest to the laboratory. If larger quantities must be kept on hand, a secured area in the building should be used. Storage for the larger quantities should be in an area where the temperature will not rise to a level that would support combustion. The nature of the liquids should also be considered, that is, large quantities of nitric acid and toluene should not be stored so close together that breakage would cause the two to mix. Storage areas should be well ventilated to prevent the buildup of possibly hazardous vapors. These areas should also not be heated or contain any electrical switches. If the flash point of the organic chemicals is below room temperature, a separate refrigerator having no internal switches for their storage should be provided. Concentrated acids should be kept in 2.2 liter bottle quantities in the stockroom and only one bottle of each acid should be kept on hand close to the laboratory. Extra quantities of these acids should be kept in a separate area as mentioned previously.

A special effort should be made to secure the stockroom during the summer months, since it is during this period that temperatures may rise and produce reactions that could proceed unnoticed until a fire breaks out. The teacher should be certain that all chemicals are in their appropriate places and that there is sufficient ventilation to keep the temperatures low enough to prevent combustion. Locks should also be provided so that unauthorized people cannot enter these areas when the school is closed. The following check list is provided as a guide to closing the laboratory and stockroom for the summer.

1. Return all pure chemicals to the stockroom.
2. Dispose of extra solutions that are not needed.
3. Wash and reshelve all glassware.
4. Remove all reagents from the laboratory to the stockroom.
5. Clean all student glassware and replace missing components.

6. Clean benchtops and clear display areas.
7. Compile the semester usage lists.
8. Remove all waste chemicals and solutions from the laboratory's preparation room and stockroom.
9. Check the labels on all bottles for tightness and legibility.
10. Inspect all bottles for cracks and/or leakage.
11. Lock the chemical storeroom door.
12. Shut off the electricity, gas, and water for the student desks and the preparation area.
13. Inform the administration and custodial personnel of where chemicals and equipment received during the summer should be stored.

Safety in the Physics Laboratory

The safety of his students is the science teacher's responsibility. He should instruct his students in safe procedures and inform them about possible bodily harm. This section contains a description of specific dangers associated with various types of equipment that might be found in the physics laboratory as well as suggestions for the safe use of this equipment.

Many accidents can be prevented by simple precautions. All equipment should be kept in good repair and safety devices should be installed wherever possible. For example, all belt-driven equipment, such as vacuum pumps, should have their belts covered. All new equipment should be given a thorough inspection and tested for possible hazards. The instruction manual should be thoroughly read. Enough room should be allowed for each group of students that is doing an experiment so that there will be no interference between groups. Students must be warned of any dangers that are peculiar to the experiment just before they begin, even if the hazards may seem obvious. Students tend to forget simple precautions such as keeping long hair away from open flames and keeping their faces away from tautly stretched cords or wires that might break.

Some hazards are less obvious and therefore must be carefully explained to the students. The proper handling of mercury is an example. Mercury is a very interesting but poisonous liquid. Even if the teacher does not allow his students to handle mercury, he will occasionally find that some mercury has escaped from broken thermometers. Mercury vapor creates a continuous cumulative hazard. Therefore,[1] all containers of mercury must be kept closed and all

---

[1] The General Safety Committee of the Manufacturing Chemists' Association, Inc., *Guide for Safety in the Chemical Laboratory*, New York: Van Nostrand Reinhold Co., Inc., 1966.

spilled mercury must be cleaned up. Mercury that cannot be completely cleaned up should be dusted with sulfur powder or zinc powder. Mercury should not come into contact with the skin, should not be heated during any classroom procedure, and should not be poured into any drain or placed with trash that will be burned.

Electrical Hazards

The reason that the shock received from the small Van de Graaff, Wimhurst or Dirod machine is not dangerous is because of the low capacitances of these machines. Even though the machine may have a potential of ten kilovolts or more, the charge transferred is small and a spark nearly discharges the machine. The energy that is transferred is important; one-quarter joule gives a heavy shock and a discharge exceeding ten joules can be hazardous.[2] The energy that is transferred can be estimated from the product of the average voltage and the charge that is transferred. It is hazardous to connect a capacitor to a Van de Graaff machine because the energy that can be transferred in a discharge is increased. Capacitors must always be handled with caution. A capacitor can become charged by a flow of ions from an electrostatic generator if the capacitor is set near the generator but not connected to it. Capacitors are seldom completely discharged by one spark because of the tendency for charge to oscillate in a capacitive system. Capacitors should therefore be shorted for longer than an instant and they should also be stored shorted.

The common 115-volt AC line should be treated with respect. Current can be continually supplied by the electric company so that energy continues to be transferred to a person as long as he is in contact with the line. The 60-cycle current can interfere with bodily functions, especially with the heart's action. As small a current as 50 milliamperes can cause a fatal shock. If the current is above about nine milliamperes, the muscles contract so that the victim cannot release his grip. If a current of twenty-five milliamperes or more passes through an individual, breathing becomes difficult or impossible.[3] In such cases the victim can be saved by removing the source of the current and applying immediate and continued artificial respiration.

This discussion does not allow sufficiently for individual differences. The physical ability of a student to withstand a shock depends upon

---

[2] A. C. Moore, *Electrostatics* (Garden City, N.Y.: Doubleday Anchor Books, 1968).
[3] "Safety in the Laboratory—Electricity," *The Physics Teacher* 7:505–506 (December, 1969).

his medical history. If a student has heart trouble or epilepsy, for example, he should not be exposed to large shocks. Furthermore, the teacher should strive to relieve fear rather than increase it; therefore he should not insist that each student experience a shock from an electrostatic generator for both medical and psychological reasons.

Grounding should be provided on all exposed metallic noncurrent carrying parts of electrical supply equipment such as frames of generators, transformer cases, switch boxes, and spark timers. Variable voltage transformers are better than a rheostat used as a voltage divider for use in alternating current demonstrations or experiments because the danger of shocks is much less. Variable voltage transformers are available from electronic supply companies for about the same price as a good rheostat purchased from a scientific supply company.

Laser Safety

Laser light must be used with caution because of its potential danger to the eye. Intense visible light can damage the retina and ultraviolet and infrared radiations can cause further damage. The following precautions are applicable to low-powered continuous-wave gas and semiconductor lasers. More precautions must be observed if other types of lasers are used.

1. Do not look into the primary beam or at mirror-like reflections of the beam. Keep all glass and other reflective surfaces away from the target area. A dark cloth drape behind the target helps to reduce reflections.

2. Avoid looking along the beam to aim the laser because this increases the hazard from reflections. It is better to use a matte-finished card to locate the beam.

3. Shatter resistant safety goggles do not permit direct viewing of the beam, although they do provide partial protection. The goggles should be designed to filter out the specific frequencies that are characteristic of *your* laser and, for a ten milliwatt/cm^2 laser, should have an optical density of about four.[4]

4. Keep the area where the laser is being used well lighted so that the pupils allow less light to enter the eye. This will slightly decrease the amount of laser light that can reach the retina if an accident should occur. If an accident should occur, the affected eye should be examined promptly by a physician.

---

[4] *A Guide for Uniform Industrial Hygiene Codes of Regulations for Laser Installations, Supplement No. 7.* The American Conference of Governmental Industrial Hygienists, 1968. Copies of this report are available for $.50 from: The American Conference of Governmental Hygienists, 1014 Broadway, Cincinnati, Ohio 45202.

5. Keep in mind that the use of external lenses and mirrors can affect the beam's power density by increasing or decreasing the size of the beam.

The maximum permissible exposure level for a continuous wave laser is about 1.0×10^{-5} watts/cm^2. Thus, for a ten milliwatt laser with a half cm diameter beam and a beam divergence of two milliradians, the safe direct viewing distance would be about 600 feet.[5]

Radiation Safety

Radiation effects on human beings are of two kinds: the body tissues may be injured or the germ plasm may be injured. Injury to body tissues may not be apparent until many years after the exposure. The doses necessary to produce such injury are, in general, considerable. For most radiation equipment likely to be found in a high school laboratory, such doses could be received by an individual only as a result of extreme carelessness or if he did not realize that radiation was present. Injury to the germ plasm of an individual might result in the transmission of hereditary defects to his descendants. Because the individual retains alteration in his genetic material, it is important to keep irradiation of the reproductive organs to a minimum.[6]

1. RADIOACTIVE SOURCES

This section is concerned with the handling and storage of only generally licensed quantities of radioactive materials. A license is required if sources of higher activity are to be used. Generally licensed quantities of radioactive materials can be sold to anyone with the understanding that they are never to be administered internally to human beings and that two or more sources are never to be combined to make a stronger source.

The dose of radiation received by an individual can be kept low by keeping the exposure time brief and by keeping the individual as far from the source of radiation as possible. Students who comprehend the "inverse square law" should understand the advantages of staying away from the source. Disk sources should be handled by their rims so that the fingers are kept as far away from the actual radioactive material as possible. Radioactive sources should never be carried in a pocket. Exposure time can be minimized by setting up all of the

---

[5] M. E. LaNier, "Evaluating a Laser Hazard," *National Safety News* (November, 1969), p. 92.
[6] *Radiation Protection in Educational Institutions.* NCRP Report No. 32. National Council on Radiation Protection and Measurements. Copies of this report can be purchased for $.75 from: NCRP Publications, P.O. Box 4867, Washington, D.C. 20008.

apparatus for an experiment or demonstration before installing the radioactive source.

Shielding should be used when the students must work near a radioactive source for long periods, if they will crowd around a table to watch a demonstration, or if an X-ray source is being used. Lead bricks are very effective for shielding and can be arranged to suit any experiment. Solid concrete bricks or ordinary bricks are less effective, but are also less expensive. For gamma or X-radiation of 1.5 Mev, three cm of concrete will decrease the beam intensity by one half, and twenty cm of concrete will decrease it by one-tenth of its incident value.

Radioactive sources must be stored in a suitable location. The teacher's desk drawer is not a suitable location because it is much too near to him. A steel file box lined with lead sheet and provided with a lock could be used to store radioactive sources. This box should be placed where students do not have access to it. The box should also be kept well away from photographic supplies such as film.

Sealed radioactive sources should be handled with care so that the seal is not broken, since the radioactive material that could escape could contaminate the air and the room. A radioactive source should not be kept if the seal is suspected of being broken. A gift of an old radium source should not be accepted, because radium sources are likely to leak and because this would probably not be a generally licensed amount. Students should not be allowed to dismantle devices such as luminous dials because of the risk of inhaling or swallowing some of the radioactive materials. The disposal of sealed sources is best done by a company that is licensed to dispose of radioactive materials. To obtain the name and address of such a company, contact the state agency that deals with radiation safety, or consult the telephone directory of a large metropolitan area.

Radioactive materials in solution require special handling. Rubber or plastic gloves should be used when transferring such solutions and the hands should be washed immediately after removing the gloves. Work should be carried out under a hood if possible, and over a metal or plastic tray that has been lined with absorbent paper. The containers should be kept closed except when some of the solution is being transferred. Pipetting by mouth should never be permitted. Inexpensive safety syringes are available to transfer the solution. Eating, drinking, or smoking in an area that has been used for experiments involving unsealed sources should be prohibited. Floors of concrete or wood are very difficult to decontaminate; such floors should therefore be covered with plastic sheets wherever spills may occur. Generally licensed radioactive solutions may be disposed of by washing them away with large quantities of water. The contaminated paper, gloves, and other contaminated articles should be disposed of in a sealed, waterproof container.

2. OTHER RADIATION SOURCES

X-ray sources should be completely shielded except for the useful beam, and the shielding should be tested for leakage. An old X-ray machine that the school may have acquired as a gift may be inadequately shielded. Unless the teacher knows how to test for leakage and how to improve the shielding if necessary, it would be best not to use the machine. No part of the human body should be exposed to the radiation beam. It is unlawful in many states to use X-rays on human beings for demonstration purposes.

X-rays can be produced by equipment designed for other purposes. Any electronic tube operating at a potential above ten kilovolts could produce X-rays, and this possibility should be investigated. High voltage sources such as induction coils and Tesla coils ordinarily do not produce X-rays, but if they are used with an evacuated discharge tube, X-rays may be produced. At voltages below ten kilovolts, the tube walls may provide adequate shielding. The amount of radiation that passes through the tube walls depends upon the current and the operating voltage. Therefore, even if the tube operates at a potential above ten kilovolts, the amount of radiation emitted from the tube will be small if the current is kept small.

The teacher must assume the responsibility for the safety of his students at all times. The teacher must be aware of the sources of potential danger and act to minimize these dangers. Careful planning and adherence to the guidelines presented in this chapter is strongly encouraged.

Science Fairs

A science fair is an exhibition of displays representing the projects of the student participants. This exhibition is usually open to the public to permit parents and other members of the community to view the project displays. Although the judging of the projects is not a necessary part of a science fair, that along with the award of prizes represents an important part of most science fairs whether the awards are in the form of ribbons or money.

A science fair should encourage student interest in science as well as participation in science activities. The evaluation of the projects and their exhibition can both be worthwhile experiences for the student exhibitors. Science fairs can also have positive public relations value for the school through the display of the projects to the parents and the community, and recognition can be given to the student participants. Although these may all be desirable, worthwhile objectives, the primary emphasis should be upon the encouragement of student interest and participation in science.

If a science fair is to emphasize science, student participation should be encouraged, but optional. When participation is required, too many student projects seem to have little or no science involved in them, and/or are not the student's own work. Required student participation may be appropriate for an honors section of a science course, but it is not a desirable procedure in general. Since the projects do require a considerable amount of the student's time, placing these activities within an independent study course or a special science seminar program and giving the student some academic credit for his work would encourage student participation in science fair activities without making it a required activity.

Suggestions for the Teacher

I. A good science fair requires much planning and effort; it is not a one-man job. Since a successful science fair can be an asset to the entire school and the community, as many of the science teachers as possible should be enlisted in the support of the science fair. If one person is to direct the fair, some released time from teaching or other duties would be desirable for this person.

II. The planning for the fair should be started *early*. This is necessary in order to provide the students enough time to select and complete their projects.

III. Arranging for the judging of the science fair may be a serious problem. Competent judges who are willing and able to devote the necessary time and effort may be difficult to find. Engineers and science personnel from industry or college and university scientists may be willing to serve as judges. However, even highly qualified judges will need judging criteria and it is best to have these clearly defined. The following criteria are used at the International Science and Engineering Fair,[1] and have been quoted with permission of Science Service, Inc.

JUDGING CRITERIA
What Are We Judging?
We are judging the following:

1. The quality of the work done on a project in science, engineering, or mathematics by a high school student and how well that student understands the project and the area in which he has been working. Only secondarily are we evaluating the physical display.

2. A project that involves laboratory, field, or theoretical work, and not just library research or gadgeteering.

3. A high school student's work, and not that of a Ph.D. candidate or a professional. Sometimes judges tend to overreact to high school students, either giving them far more credit than they deserve or acting as though the work done by the student was worthless because it was not in the Nobel Prize category.

4. A project as compared with the other projects in the same category and not with projects seen elsewhere under other circumstances.

Criteria

Exhibits are judged on the following basis:

Creative ability	30 points
Scientific thought/Engineering goals	30 points
Thoroughness	10 points
Skill	10 points
Clarity	10 points
Dramatic value	10 points

---

[1] Science Service, Inc., *Judging Guide 22nd International Science and Engineering Fair* (Washington, D.C.: Science Service, Inc., 1971), pp. 7–11.

Creative Ability (30 points)
1. Does the project show creative ability and originality in
    —the question asked?
    —the approach to solving the problem?
    —the analysis of the data?
    —the interpretation of the data?
    —the use of equipment?
    —the construction or design of new equipment?

Obviously, no project would be creative and original in all these aspects, and in addition one must keep in mind that one is dealing with high school students. Thus, one must ask whether something is creative and original in terms of a professional level, or for a high school student. The latter is most probable, and means that it is very important to try to ascertain the nature of the assistance that the student has received.

A student should not be penalized for taking help from others (all professionals receive help to some degree in some way). But credit for creative ability and originality should be in regard to what the student himself has contributed, and not for what others have done for him.

For example, did a student get an idea for his project from a textbook suggestion for research, or did he develop the idea himself as a result of reading or work that he had done? If he developed the idea himself, it would be considered more creative.

A warning to judges should be made at this point. There have been projects that had elements in them that judges thought were original, but that actually came out of textbooks or laboratory manuals in newly developed curricula with which they were unfamiliar. This possibility should be kept in mind.

Another source of help that should be evaluated is that received from a teacher or other adult. A student may have a very original approach for solving a problem, but it may have come out of suggestions made by a scientist or engineer with whom he worked during the summer. This idea must be compared with something less sophisticated, but that which came genuinely from the work or thinking of a student. The latter would be considered more creative.

2. Collections cannot be considered to be creative unless they are used to support an investigation and help to answer a question in some original way. Construction of equipment that involves the assembly of a kit cannot be considered to be creative unless some unusual approach or design is used.

3. For engineering, a clear distinction should be made between gadgeteering and a genuine contribution. A "Rube Goldberg" device may be ingenious, but if it is not really the most efficient way to solve a problem, if it is not acceptable to the potential user, if it is unreliable in its functioning, then it cannot really be considered to be a valuable creative contribution.

Scientific Thought/Engineering Goals (30 points)
Scientific Thought:
1. Is the problem stated clearly and unambiguously?

2. Is the problem sufficiently limited so that it is possible to attack? One of the characteristics of good scientists has been reported to be the ability to identify important problems that are capable of solution. Simply working on a difficult problem without getting anywhere does not make much of a contribution. On the other hand, neither does solving a very simple problem.
3. Was there a procedural plan for obtaining a solution?
4. Are the variables clearly recognized and defined?
5. If controls were necessary, was there a recognition of their need and were they correctly used?
6. Are there adequate data to support the conclusions?
7. Are the limitations of the data recognized?
8. Does the student understand how his project ties in with related research?
9. Does the student have an idea of what further research is indicated?
10. Did the student cite scientific literature, or did he cite only popular literature (local newspaper, *Reader's Digest*)? Note: It should be pointed out again that the student may have received assistance, and that it is important to estimate the extent of this assistance, and what contribution it made to the project.

Engineering Goals:

Note: We have not had much experience in applying these criteria to student projects, and so judges are urged to use them as guide lines only. Remember, we are judging among the exhibits in the ISEF, and not against engineering projects done by professional engineers.
1. Does the project have a clear objective?
2. Does this objective have relevance to the needs of the potential user?
3. Is the solution
   —workable? Unworkable solutions may be interesting but are of no value from a practical point of view.
   —acceptable to the potential user? Solutions that will be rejected or ignored are of no value.
   —economically feasible? A solution that is so expensive that it cannot be utilized is of no value.
4. Can the solution be successfully utilized in design or construction of some end product?
5. Does the solution represent a significant improvement over previous alternatives?
6. Has the solution been tested to see if it will perform under the conditions of use? (This may be difficult for many students, but it should at least be considered.)

Thoroughness (10 points)

1. Does the project carry out its purposes to completion within the scope of the original aims?
2. How completely has the problem been covered in the project?
3. Are the conclusions based on a single experiment or on replication?

4. If it is the kind of project where notes were appropriate, how complete are they?
5. Is the student aware of other approaches or theories concerning his project?
6. How much time was spent on the project?
7. Is the student familiar with the scientific literature in the field in which he was working? Note: Citations are not considered to be an important consideration in engineering (as opposed to science) and so a student should not be penalized for a lack of citations.

Skill (10 points)

1. Does the student himself have the skills required to do all the work necessary to obtain the data that support his project, i.e., laboratory skills, computation skills, observational skills, and design skills?
2. Where was the project done—at home, in the school laboratory, or in the university laboratory? What assistance was received from parents, teachers, scientists, or engineers?
3. Was the project carried out under the supervision of an adult or did the student work largely on his own?
4. Where did the equipment come from? Did the student build it himself? Was it obtained on loan? Was it part of a laboratory in which he worked?

Clarity (10 points)

1. How clearly is the student able to discuss the project? Is he able to explain its purpose, procedure, and conclusions in a clear and concise manner? Discount a glib tongue, but try to make allowances for nervousness that may result from talking to an authority. Try to watch out for memorized speeches with little understanding of principles.
2. Has the student expressed himself well in written material? Remember that such material could have been prepared with the assistance of another person.
3. Are the important phases of the project presented in an orderly manner?
4. How clearly are the data presented?
5. How clearly are the results presented?
6. How well does the project display explain itself?

Dramatic Value (10 points)

1. How attractive is the exhibit compared to the others? A working exhibit should not be given any more credit than a nonworking one.
2. Is the presentation done in a forthright manner, without cute tricks or gadgets?
3. Was all the work done by the student or did he receive assistance from his art class or others?

IV. The preceding judging criteria are designed for science projects based upon experiments. Exhibits that represent only collections, models, or constructed apparatus such as telescopes are not in themselves experimental projects. These exhibits may represent a very worthwhile activity for the students, but it can be asked, "Are they really science?" (LaNeve, 1967) If these exhibits are included in a science fair, a separate judging category should be provided for them.

Suggestions for the Student

1. Define your problem clearly. This will make the work on the project easier and it will be extremely important when the project is being judged.
2. Get started on the project early. Establish some kind of a regular work schedule and stick to it.
3. Keep a notebook of your work on the project. This should be a complete record of your observations, inferences, and your data for both successful and unsuccessful parts of the investigation. This notebook will be the basis for your project report.
4. Use your local information sources first. Both school and public libraries have science books and magazines as well as general references such as encyclopedias. These sources will provide enough information to get you started on a project. Be sure that you thoroughly utilize these resources before asking for assistance from outside of your school. A letter that states, "Send everything you have on pollution." is unlikely to result in anything more than a letter directing you to your local library. You are more likely to get help if you have your problem clearly defined and can ask specific questions.
5. The AEC pamphlet *Atoms at the Science Fair* gives useful directions for the construction of a project display.

REFERENCES

Bacon, W. S., ed., *Ideas for Science Fair Projects* (New York: Fawcett Publications, Inc., 1962), 144 pages, $.75. Suggestions from the winners of the National Science Fair-International.

Chesley, Robert E., "Student Projects in Physics," *The Physics Teacher*, 7:449-451 (1962). After a brief, general discussion of good and bad projects, sixteen specific ideas suitable for student projects in physics are presented.

LaNeve, Edward B., "Are Science Fairs *Really* Science?" *The Science Teacher*, 34(9):56 (December, 1967). An essay critical of projects that are not representative of scientific work but that often win prizes at science fairs.

LeCompte, Robert G., and Burrell L. Wood, *Atoms at the Science Fair* (Oak Ridge, Tenn.: United States Atomic Energy Commission, 1968), 55 pages, Free. This booklet provides a comprehensive discussion of science projects and how to display them at science fairs. This booklet is oriented toward the student-participant and provides lists of possible topics along with advice on everything from the choice of the project to the construction of the display for the completed project. Available from: USAEC, P. O. Box 62, Oak Ridge, Tennessee, 37830.

National Science Teachers Association. Information on student groups, such as Future Scientists of America, and national activities such as the Youth Science Congresses, meetings sponsored by NASA and the NSTA where students can report on their experimental investigations and interact with other students and scientists, can be obtained from: NSTA, Mrs. D. K. Culbert, Assistant Executive Secretary for Student Programs, 1201 Sixteenth Street N.W., Washington, D.C. 20036.

Science Publications. This publisher has several books that could be useful references for students participating in science fairs or science club activities. A list of these books can be obtained from: Science Publications, P. O. Box 286, Normal, Illinois, 61761.

Shay, Edwin L., "Science Fairs, not Contests," *The Science Teacher*, 35(4): 55-56 (April, 1968). A reply to LaNeve's essay suggesting an interesting alternative to judged science fairs.

Showalter, Victor, and Slesnick, Irwin, *Award Winning Future Scientists of America Science Projects* (Washington, D.C.: National Science Teachers Association), 30 pages, $1.00. Abstracts of winning FSA projects. These can be used as a possible source of ideas for projects and as examples of what goes into a good project.

Welte, Dimond, and Friedl, *Your Science Fair*, 3rd ed. (Minneapolis, Minn.: Burgess Publishing Co., 1966), 166 pp., $3.25. A guide for the science teacher with little or no experience with science fairs. It provides suggestions for the organization and execution of a science fair.

Periodicals and Organizations

Periodicals

All that changes is constant and all that is constant changes. With this in mind it is necessary to consider the function of periodicals with respect *to up-dating the teacher* and opening new horizons for the students.

A library that embraces a large collection of magazines provides the student with an opportunity to become aware of the many diverse events occurring in science. The student is made aware of what is being accomplished in the research laboratory and how this research is being translated into goods and services.

The science teacher has the obligation of insuring that the library subscribes to periodicals that are directly related to his teaching discipline. As a further criterion, the teacher should select only those periodicals that are interesting to and on the reading level of the high school student. Research journals do not typically appeal to a sufficient number of students to warrant their inclusion in the high school library.

Once the teacher has made the transition from the role of student to the role of teacher and the years pass, the information and the concepts used by the teacher in the classroom must be evaluated. This evaluation takes the form of reviewing what is being taught in the classroom with respect to what is evolving in the university research laboratories and the applications of the new ideas by industry. The teacher has two options in this evaluation. He may either choose to return to school on a periodic basis or he may elect to stay abreast of his field by the use of journals and other periodicals. It is recommended that both methods be used. The teacher should also remember that the face of education is changing and that he will also need to be informed of changes in education.

The list of publications that follows is intended to be used as a guide by the teacher in determining the composition of the library's periodical collection in addition to his private collection. The numbers used will indicate which category or objective the periodical can serve for the teacher. A periodical marked "1" is recommended for the school library, one marked "2" is recommended for the teacher's library, and one marked "3" is recommended for inclusion in the student's library.

The following magazines are recommended as noted for either the school library or the teacher's library. The cost of each subscription and a description of the type of materials that may be found in each publication are provided. The teacher is encouraged to read as many of these magazines as possible to provide him with the current diversified background that is required to make the concepts to be taught relevant to the progress being made by the technologically based industries.

The American Chemical Society publishes many varied journals to appeal to the special interests of the chemical community. The teacher of chemistry is urged to read and/or subscribe to any of these journals to enable him to stay abreast of research developments. The following is a list of the periodicals offered by this organization and the cost of each. Since prices are subject to change, the dollar amounts shown here are intended to provide the reader with an indication of the relative cost of the items. They may be ordered by writing to: The American Chemical Society, 1155 Sixteenth St., N.W., Washington, D.C. 20036.

		Nonmember	*Member*
1.	Chemical Technology[2]	$18.00	
2.	Analytical Chemistry[2]	$ 7.00	$ 5.00
3.	The Journal of the American Chemical Society[2]	$44.00	$22.00
4.	Biochemistry[2]	$20.00	$20.00
5.	The Journal of Medicinal Chemistry[2]	$30.00	$15.00
6.	Inorganic Chemistry[2]	$36.00	$18.00
7.	Journal of Organic Chemistry[2]	$40.00	$20.00
8.	Chemical and Engineering News[2]	$ 7.00	Subscription with Membership
9.	Journal of Agricultural and Food Chemistry[2]	$20.00	$10.00
10.	Journal of Chemical and Engineering Data[2]	$30.00	$15.00
11.	Macromolecules[2]	$24.00	$12.00
12.	Accounts of Chemical Research[2]	$10.00	$ 5.00
13.	Chemical Reviews[2]	$26.00	$13.00
14.	The Journal of Physical Chemistry[2]	$40.00	$20.00

	Nonmember	Member
15. Journal of Chemical Documentation[2]	$14.00	$ 7.00
16. Environmental Science and Technology[2]	$10.00	$ 5.00
17. Journal of Chemical Education[2]	$ 6.00	
18. Chemistry[2]	$ 6.00	

American Journal of Physics[2]

American Institute of Physics
335 E. 45th Street
New York, New York 10017

1 year $18.00. May be selected by members of the American Association of Physics Teachers as one of the Journals supplied with membership in the organization; annual dues are only $15.00. Published monthly. This journal is intended for use by college and university physics teachers. It is primarily devoted to articles dealing with the teaching of physics and research at the college level.

Chemistry[1,2,3]

Published by the American Chemical Society, monthly, except a combined July/August issue. 1 year, $6.00. This magazine will introduce the student to the field of chemistry through a series of articles covering current and historical topics. Experiments are included that may be used by the students in the laboratory. Short news notes are included to keep the student and the teacher informed on the latest events in chemistry. Class orders are available at a reduced rate when ordered in lot of ten or more.

Education in Chemistry[2]

Chemical Society Publications Sales Office
Blackhorse Road
Letchworth, Herts
England

1 year $10.60. Published January and alternate months. This publication contains news of progress made in teaching chemistry in the

United Kingdom. It is useful for comparative purposes and provides a different view of the educational process than what is normally seen in American journals.

Electronics World[1]

Circulation Department
P. O. Box 1093
Flushing, New York 11352

1 year, $7.00. Published monthly. This magazine provides popular product evaluations and articles for special projects that could be attempted by students.

Environmental Science and Technology[1]

Subscription Services Department
American Chemical Society
1155 Sixteenth Street, N.W.
Washington, D.C. 20036

1 year, $5.00. Published monthly. This publication may be used to keep abreast of the current trends in activities that relate to the environment.

Journal of Chemical Education[1,2]

Business and Publications Office
20th and Northampton Streets
Easton, Pa., 18042

1 year, $6.00. Published monthly. This journal is oriented to the teacher of chemistry with articles on the latest research in chemistry that is applicable to the classroom. In addition, there are articles detailing successful experiments in the teaching of chemistry as well as experiments that have proven successful at other institutions.

Journal of Research in Science Teaching[2]

Journal of Research in Science Teaching
605 Third Avenue
New York, New York 10016

1 year, $10.00. Published quarterly. This is the official journal of the National Association for Research in Science Teaching. It pre-

sents reports of research studies in science education and is oriented toward the science educator.

Nature[1]

MacMillan Journals Ltd.
711 National Press Building
Washington, D.C. 20004

Published weekly. This is one of the oldest journals in science giving broad coverage to all aspects of science and the research being conducted to advance general knowledge in the field of science.

Physics Education[2]

American Institute of Physics
335 East 45th Street
New York, New York 10017   (for North American orders)

1 year, $14.00. Published bi-monthly. This journal is concerned with physics teaching in Great Britain and the United Kingdom. Its content and level of presentation are very similar to *The Physics Teacher*, which is published in the United States.

The Physics Teacher[2]

American Institute of Physics
335 East 45th Street
New York, New York 10017

1 year, $8.00. Published monthly September through May. This journal may be selected as the membership journal for members of the American Association of Physics Teachers. Annual dues are $15.00; members also receive *Physics Today*. An excellent publication for the high school physics teacher. Articles include science history, physics teaching, and topics of current interest in physics. A unique feature of this journal is its periodic comparative analyses of various manufacturer's versions of apparatus such as ripple tanks or electroscopes. Short notes on apparatus, student experiments, and teaching are also included. The book review section would also be of considerable interest to the physics teacher.

Physics Today[2]

American Institute of Physics
335 East 45th Street
New York, New York 10017

1 year, $7.00. Published monthly. This journal also comes with membership in any of the member societies of the American Institute of Physics, such as the American Association of Physics Teachers. This magazine provides fairly sophisticated review articles on current topics in physics. News and items of interest such as graduate enrollments in physics and employment of physicists are also discussed.

Popular Science[1,2,3]

Popular Science
Subscription Department
Boulder, Colorado 80302

1 year, $5.00; 2 years, $9.00; and 3 years, $12.00. Published monthly. This magazine offers a variety of short news articles and feature articles on a variety of science topics that relate to the improvement of the general physical well-being of the individual. It is one of the more popular nontechnical science magazines on the market.

School Science and Mathematics[2]

School Science and Mathematics Association, Inc.
P. O. Box 246
Bloomington, Indiana 47401

Subscription comes with membership in the Association. Published monthly except July, August, and September. The articles in this journal span the elementary and secondary school curriculum in mathematics and science. Topics include research in the teaching of these areas and information on innovative procedures.

Science[1]

American Association for the Advancement of Science
1515 Massachusetts Avenue, N.W.
Washington, D.C. 20005

Subscription comes with membership in the organization. Published weekly. The entire area of science developments is surveyed in the articles of this journal. It is probably more highly research oriented than most of the other journals.

Science and Children[2]

National Science Teachers Association
1201 Sixteenth Street, N.W.
Washington, D.C. 20036

1 year, $4.00. Published monthly September through December and February through May. This journal provides many ideas for teaching different concepts of science to elementary students. It is written to aid the elementary school teacher in implementing a science program that is understandable by the students.

Science Digest[1,2,3]

Science Digest
Box 654
New York, New York 10019

1 year, $5.00. Published monthly. This is one of the more popular magazines for student use. The articles are written on the format of *Reader's Digest.* Recent news from research and how it is being applied by industry is reported. In addition, it contains news about what is being done with respect to the improvement of the environment.

Science Education[2]

Science Education
University of Tampa
Tampa, Florida

1 year, $5.00. Published five times a year. The articles in this journal pertain to the teaching of science and reports the results of the research conducted with the objective of improving science teaching.

The Science Teacher[2]

National Science Teachers Association
1201 Sixteenth Street, N.W.
Washington, D.C. 20036

Subscription comes with membership in the Association. Published September through May. This periodical contains articles of interest to the science teacher on the trends of science education and educational research being conducted to advance the teaching of science.

Scientific American[1,2,3]

Scientific American
415 Madison Avenue
New York, New York 10017

1 year, $10.00; 2 years, $18.00; and 3 years, $25.00. This is probably the best magazine for keeping a science teacher current with the latest events in every field of science. The articles contain information that will provide background information as well as updating previous discoveries. This magazine is a must for every teacher's library as well as the school's library. Articles from this magazine have inspired many science projects. A special section is included that contains information for building inexpensive equipment for conducting research or studying effects.

    The teacher's responsibility continues beyond having the magazines available for the students in the school or classroom library. The teacher must make a concerted effort to involve each student with magazines that he would otherwise ignore. Blanket assignments to read some magazine once in a while is not enough exposure for the students. A more directed approach will show the students the advantages of reading science magazines for the articles that will make a difference in their lives. Magazines also serve the valuable purpose of updating the textbook being used and introducing material from the other science disciplines that might otherwise be overlooked.
    Magazines such as *Chemistry* frequently publish articles on the history of chemistry. These articles deal with the personalities of the individuals most closely associated with the given concepts and give the student a humanistic view of the subject.
    Some of the magazines include laboratory experiments that may be conducted in the high school laboratories. These may either be repetitions of the classical experiments that led to the initial discoveries or a new experiment that is a synthesis of one or more older ones. These experiments usually include a reading list that provides the background for the concept to be studied. This enables the students to gain an understanding of the amount of preliminary work that must be accomplished before proceeding to the work in the laboratory.
    The teacher has several options available to him in the use of this supplementary resource. Many teachers tend to use magazines infrequently and then only for special occasions. This is not as desirable as a regular magazine study program. It is possible with weekly periodicals to spend one day each week reading magazines. Since the

majority of the periodicals are monthlies, it may be more profitable to use one or two consecutive days each month looking through the magazines for specific topics. Students could be encouraged to browse through a number of periodicals and record articles in a particular area in a running bibliography that may be connected with a term paper or independent study. If possible, the teacher might use the school's facilities to duplicate articles for future reference.

The continued and increasing emphasis on environmental quality and control demands that the science teacher and student be as knowledgeable as possible in this area. This may be accomplished most effectively by using the periodicals of the school library to the maximum possible extent. If the student has a good background of actual experiences, either directly or indirectly, he may be able to learn concepts more readily than if the concepts were presented to him without some experience in the physical world to which the student could relate. The teacher must insure that the student know how to use the periodicals as well as which periodicals provide the most information.

Organizations

Professional organizations are available to assist the new teacher. The usual question asked by the teacher is: "Why should I join?" The answer is not always a simple one. If the teacher does not want to expend any effort in the exchange of ideas in either the discipline or in teaching, there will probably be no gain to the teacher in joining an organization. This type of individual will usually try to keep current in the field by occasionally reading a journal. If the teacher desires to be a part of both the field of science and of education, he will want to become involved with people who are actively working in these areas. Professional organizations provide the service of a clearinghouse for ideas and a medium through which these ideas can be exchanged. Teachers with a common objective and common problems meet through these organizations to seek the most viable solutions to these problems.

The teacher must be advised that many different types of professional organizations can serve his needs and should join only the ones that will benefit him most. Two basic divisions can be made in the professional organizations that will be of interest to the teacher: chemistry and physics organizations and education organizations. Within these divisions the teacher will have a selection among national, state, and local organizations.

The American Chemical Society and the American Association of Physics Teachers are the two organizations related to the teaching discipline. The orientation of these organizations is usually toward the latest developments in research in the field. However, each organization has divisions that should reflect the teaching interests of its members. Experience indicates that high school teachers of chemistry or physics are not actively involved in determining the objectives of either of these organizations or their affiliates. The goals that have been set in the past reflect this observation. Each organization publishes a variety of journals that will provide most of the service to the chemistry or physics teacher.

The education organizations range from those interested in general education to those interested in science education. The teacher has only a limited amount of time and money to spend on organizations and the teacher should become involved with only those organizations that are involved with both science and the teaching of science. The professional organizations in this area include the National Science Teachers Association, the Central Association of Science and Mathematics Teachers, and the National Association for Research in Science Teaching. Of these the new teacher will probably receive the most value from joining the National Science Teachers Association.

In addition to their publications, each of these organizations holds national meetings at least once a year. These meetings are highlighted by outstanding speakers in the fields of science and education. In addition the member has the opportunity to participate in many small interest group sessions in which innovations are discussed and demonstrated and specific problem areas in science teaching are examined. Teachers attending such meetings usually return to their schools with a variety of new ideas.

It is not always possible to attend the meetings of these national organizations. The teacher should then look to state and local organizations in the same general area. Many states have associations of chemistry teachers and physics teachers in addition to science teachers as a whole. These meetings are more responsive to the needs and interests of a particular geographic area, and the relatively low cost of attending these meetings will not prohibit the teacher from participating. These state and local organizations provide a greater opportunity for the teacher to become involved in deciding objectives and in implementing them.

Numerous teachers choose not to become involved in professional organizations because they do not want to spend the time or money and are content to teach as they were taught throughout their entire career. A person who has chosen to teach in the area of chemistry or physics should at least objectively evaluate each of the organizations discussed with a trial membership. The continuation of his membership should then be dependent on the services provided by the organization.

The dues charged and the services rendered vary with each organization. If the teacher desires to learn what each has to offer, he should write to each at the following addresses:

CHEMISTRY

American Chemical Society
1155 Sixteenth Street, N.W.
Washington, D.C. 20036

PHYSICS

American Association of Physics Teachers
335 East 45th Street
New York, New York 10017

SCIENCE EDUCATION

National Science Teachers Association
1201 Sixteenth Street, N.W.
Washington, D.C. 20036

Central Association of Science and Mathematics Teachers
P.O. Box 246
Bloomington, Indiana 47401

National Association for Research in Science Teaching
Dr. T. Wayne Taylor
Science and Mathematics Teaching Center
Michigan State University
East Lansing, Michigan 48823

Library Resources

The neophyte teacher and the newly assigned teacher will need to know what library resources are available to him to complement his classroom instruction. The librarian or the science department will hopefully have a complete inventory of the chemistry and physics books on hand as well as the books of general interest to science students and teachers. Each teacher is familiar with a certain number of books from his course work or serious browsing through periodicals and other libraries. Unfortunately, in the field of science new and revised books are produced daily reflecting changes in the field. A library that was considered adequate five years ago does not have the capabilities of meeting today's requirements unless volumes have been added to it on a regular basis. Teachers no longer rely on the standard textbook as the sole source of information for their students. A library having a diversity of books on science will provide one way for a teacher to individualize his instruction.

Before attempting to recommend the purchase of books, the teacher must be informed of the current resources. A complete inventory of science books in the laboratories as well as in the library should be noted for future use. The teacher can then compare this list to the recommended books found in this chapter.

The performance objectives of the teacher's course will determine those books that will facilitate the instruction of the subject matter. With book budgets now often being pared to a minimum, the teacher should limit the books he is considering for purchase to those directly suited for student use. A good library, rather than merely having a given number of volumes in a given subject matter area, has regular student use of the volumes that are available.

Books should be ordered that will broaden a student's background in chemistry or physics or enable him to eliminate his deficiencies in these areas. The teacher should avoid the use of chemistry and physics books associated with subject matter courses of third year college level or higher. Typically the student has neither the mathematical maturity or the desire to digest the theories found in such books. Rather, he is interested in specific areas on a qualitative basis. If the teacher feels that he has a personal need for particular books, he should either purchase the necessary books or use those available at the nearest college or university library. The most valued books for the student are reviewed in *Chemistry, Journal of Chemical Edu-*

cation, *Physics Today*, *The Physics Teacher*, *The Science Teacher*, and *Scientific American*; these reviews should help the teacher determine relative effectiveness of books for use in the teacher's own course.

The amount of money that will be available to the teacher during his first year in the system will not be large, and it will be necessary for him to struggle along until a convincing argument can be presented to either his department chairman or principal. The teacher who determines in advance the need for each book and can describe how the book will aid him in his course will have the greatest likelihood of securing the money necessary to purchase new books.

The library will be used only if the students find it interesting or valuable as an aid for success in their science courses. Students will generally gravitate to the periodicals and the science fiction books before they will study a chemistry or physics book for pleasure. The teacher must introduce the students to the books in the library by requiring them to use these books in the solution of class problems. References cited in the experiments should be examined by the students and included in the write-up of the experiment if possible. Historical papers assigned in class will lead the student through several volumes and will possibly instill in him the excitement that gripped the original experimenters.

Other types of projects will bring the student in contact with the books on the shelf. The books will receive further use if they are also available in the classroom. A reference shelf should be established in each science room, which should contain other textbooks for the high school chemistry and physics courses, mathematical review books, and reference books of physical data. The teacher should also make five or more selections from the library every two weeks. These selections should be brought to the students' attention by bulletin board displays or assignment. Time should be made available to the student to browse through these books. If there is study space available in the classroom, some of the programmed texts that are now available should be kept ready for the student who falls behind in his classwork or for the interested student who desires to explore beyond the topics included in the classroom assignments.

The bibliography that is included in this chapter is designed to acquaint the teacher with the books that are available in a number of areas. It is not possible to include all of the books that would be of interest to the teacher or student. The books listed have been found helpful by various teachers, and it is hoped that they will serve as a core for a library that is usable by the student. Other books are included that only will be of interest to the teacher as a matter of personal growth or as a resource to his science teaching. Space is provided at the end of each major heading so that the teacher can update the list as he browses through various periodicals in science and education.

Books for the Chemistry Library

ANALYTICAL CHEMISTRY

Ayres, Gilbert H., *Quantitative Chemical Analysis*, 2d ed. (New York: Harper & Row, Publishers, 1968), $11.95.

Bates, Roger G., *Determination of pH; Theory and Practice* (New York: John Wiley & Sons, Inc., 1964), $13.95.

Bible, Roy H., Jr., *Guide to the NMR Empirical Method*; a workbook (New York, Plenum Publishing Corporation, 1967), $9.50.

Blaedel, Walter J., and Vulliers W. Meloche, *Elementary Quantitative Analysis, Theory and Practice*, 2d ed. (New York: Harper & Row, Publishers, 1963), $11.95.

Bobbitt, James M., Arthur S. Schwarting, and Roy J. Gritter, *Introduction to Chromatography* (New York: Van Nostrand Reinhold Company, 1968), Reinhold Science Studies, $4.50 pb.

Bovey, Frank A., *Nuclear Magnetic Resonance Spectroscopy* (New York: Academic Press, Inc., 1969).

Brumblay, Ray U., *Quantitative Analysis* (Reading, Mass.: Addison-Wesley Publishing Company, Inc., 1970).

Christian, Gary D., *Analytical Chemistry* (Boston: Ginn and Company, 1971).

Crow, D. R., and J. V. Westwood, *Polarography* (New York: Barnes & Noble, Inc., 1968).

Dean, John A., *Chemical Separation Methods* (New York: Van Nostrand Reinhold Company, 1969).

Ewing, Galen W., *Instrumental Methods of Chemical Analysis*, 3rd ed. (New York: McGraw-Hill Book Company, Inc., 1969), $11.50.

Fischer, Robert B., and Dennis G. Peters, *Basic Theory and Practice of Quantitative Chemical Analysis*, 3rd ed. (Philadelphia: W. B. Saunders Company, 1968), $11.75.

Fischer, Robert B., and Dennis G. Peters, *A Brief Introduction to Quantitative Chemical Analysis* (Philadelphia: W. B. Saunders Company, 1969), 537 pp., $9.25.

Flaschka, H. A., A. J. Barnard, and P. E. Sturrock, *Quantitative Analytical Chemistry, Volume 1: Introduction to Principles* (New York: Barnes & Noble, Inc., 1969).

Flaschka, H. A., A. J. Barnard, and P. E. Sturrock, *Quantitative Analytical Chemistry, Volume 2: A Short Introduction to Practice* (New York: Barnes & Noble, Inc., 1969).

Hahn, Richard B., and Frank J. Welcher, *Inorganic Qualitative Analysis*, 2d ed. (New York: Van Nostrand Reinhold Company, 1968), 326 pp., $6.40.

Heftmann, Erich, ed., *Chromatography*, 2d ed. (New York: Van Nostrand Reinhold Company, 1967), Reinhold chemistry textbook series, $27.50.

King, Edward J., *Qualitative Analysis and Electrolytic Solutions*, under the general editorship of Larkin H. Farinholt (New York: Harcourt Brace Jovanovich, Inc., 1959), $9.95.
Laitinen, Herbert A., *Chemical Analysis* (New York: McGraw-Hill Book Company, Inc., 1960). McGraw-Hill series in advanced chemistry, $14.50.
Malmstadt, Howard V., and Christie G. Enke, *Electronics for Scientists: Principles and Experiments for Those Who Use Instruments*. With the assistance of E. C. Toren, Jr. (Menlo Park, Calif.: W. A. Benjamin, Inc., 1962), $14.00.
Noyes, W. Albert, Jr., George S. Hammond, and J. N. Pitts, Jr., eds., *Advances in Photochemistry, Volume 6* (New York: Interscience Publishers, John Wiley & Sons, Inc., 1968), 484 pp., $19.00.
Pecsok, R. L., and L. Donald Shields, *Modern Methods of Chemical Analysis* (New York: John Wiley & Sons, Inc., 1968), 480 pp., $9.95.
Reilley, Charles N., and Donald T. Sawyer, *Experiments for Instrumental Methods: A Laboratory Manual* (New York: McGraw-Hill Book Company, Inc., 1961), $7.50.
Skoog, Douglas A., and Donald M. West, *Fundamentals of Analytical Chemistry*, 2d ed. (New York: Holt, Rinehart and Winston, Inc., 1969), $13.00.
Stahl, Egon, ed., *Thin Layer Chromatography: A Laboratory Handbook*, 2d ed. (New York: Springer-Verlag New York, Inc., 1969), 1041 pp., $32.
Willard, Hobart H., Lynne L. Merritt, Jr., and John A. Dean, *Instrumental Methods of Analysis*, 4th ed. (New York: Van Nostrand Reinhold Company, 1965), $11.50.

BIOCHEMISTRY

Calvin, Melvin, *Chemical Evolution: Molecular Evolution Towards the Origin of Living Systems on the Earth and Elsewhere* (New York: Oxford University Press, 1969), 278 pp., $4.50.
Cheldelin, Vernon H., and R. W. Newburgh, *The Chemistry of Some Life Processes* (New York: Van Nostrand Reinhold Company, 1964), Selected topics in modern chemistry, $2.25 pb.
Christensen, Halvor N., *Body Fluids and the Acid-base Balance: A Learning Program for Students of the Biological and Medical Sciences* (Philadelphia: W. B. Saunders Company, 1964), $6.50 pb.
Gutfreund, H., *An Introduction to the Study of Enzymes* (New York: John Wiley & Sons, Inc., 1965), $9.95.
Harrison, Kenneth, *A Guide-book to Biochemistry*, 2d ed. (New York: Cambridge University Press, 1965), $3.95; $1.95 pb.
Mahler, Henry R., and Eugene H. Cordes, *Biological Chemistry* (New York: Harper & Row, Publishers, 1966), $17.50.
Watson, James D., *Molecular Biology of the Gene* (Menlo Park, Calif.: W. A. Benjamin, Inc., 1965), Biology teaching monograph series, $12.50; $6.95 pb.

CHEMISTRY HISTORY

Burland, C. A., *The Arts of the Alchemists* (New York: The Macmillan Company, 1968), 244 pp., $9.95.

Conant, James B., *On Understanding Science; An Historical Approach* (New Haven: Yale University Press, 1947), The Terry Lectures, $2.00.
Faraday, Michael, *Chemical History of a Candle*, edited by William Crookes. With original illustrations and a new introduction by L. Pearce Williams (New York: Collier Books, Inc., 1962), $.95 pb.
Farber, Eduard, *The Evolution of Chemistry: A History of Its Ideas, Methods and Materials*, 2d ed. (New York: The Ronald Press Company, 1969), 437 pp., $10.00.
———, *Milestones of Modern Chemistry: Original Reports of the Discoveries* (New York: Basic Books, Inc., 1966), $5.95, Science and discovery books.
Hoffman, Banesh, *The Strange Story of the Quantum: An Account for the General Reader of the Growth of Ideas Underlying Our Present Atomic Knowledge*, 2d ed. (New York: Dover Publications, Inc., 1959), $2.00 pb.
Ihde, Aaron, J., and William F. Kieffer, *Journal of Chemical Education*, Selected readings in the history of chemistry (Easton, Pa.: Chemical Educational Publishing Company, 1965), $4.50.
Jaffe, Bernard, *Crucibles: The Story of Chemistry from Ancient Alchemy to Nuclear Fission*, newly rev. (New York: Fawcett World Library, A Premier book, 1960), $.60 pb.
Weeks, Mary E., *Discovery of the Elements*, Completely rev. and new material added by Henry M. Leicester, Illus. collected by F. B. Dains, 7th ed. (Easton, Pa.: Chemical Education Publishing Company, 1968), $12.50.

CHEMISTRY PROBLEMS AND MATHEMATICAL SKILLS
Belcher, J., J. Colbert, and H. H. Rowley, *Properties and Numerical Relationships of the Compounds* (New York: Appleton-Century-Crofts, 1962).
Frey, Paul, *Chemistry Problems and How to Solve Them*, 7th ed., (New York: Barnes and Noble, Inc., 1969), 288 pp., $1.95.
Garrett, A. B., and Arthur Kiefer, *Mathematics in Chemistry* (Boston: Ginn and Company, 1961).
Masterton, William and Emil Slowinski, *Mathematical Preparation for General Chemistry* (Philadelphia: W. B. Saunders Company, 1970).
O'Malley, Robert, *Problems in Chemistry* (New York: McGraw-Hill Book Company, Inc., 1968), 288 pp., $3.95.
Peters, Edward I., *Problem Solving for Chemistry* (Philadelphia: W. B. Saunders Company, 1971).
Pierce, Conway, and R. Nelson Smith, *General Chemistry Workbook* (San Francisco: W. H. Freeman and Company, Publishers, 1971).
Routh, Joseph I., *Mathematical Preparation for Lab Technicians* (Philadelphia: W. B. Saunders Company, 1971).
Schaum, Daniel, and Jerome L. Rosenberg, *College Chemistry*, 5th ed. (New York: McGraw-Hill Book Company, 1966).
Sienko, Michell J., *Freshman Chemistry Problems and How to Solve Them. Part I: Stoichiometry and Structure. Part II: Equilibrium* (Menlo Park, Calif.: W. A. Benjamin, Inc., 1964), $2.95 ea. pb.
Sorum, C. H., *How to Solve General Chemistry Problems*, 4th ed. (Englewood Cliffs, N.J.: Prentice-Hall, Inc., 1969), 320 pp., $3.95.

INORGANIC CHEMISTRY

Basolo, Fred, and Ronald C. Johnson, *Coordination Chemistry: The Chemistry of Metal Complexes* (Menlo Park, Calif.: W. A. Benjamin, Inc., 1964), The general chemistry monograph series, $6.00; $2.95 pb.

Bauer, Max, *Precious Stones: A Popular Account of Their Characters, Occurrence, and Applications* (Rutland, Vt.: Charles E. Tuttle Co., Inc., 1968), 647 pp., $17.50.

Chedd, Graham, *Half-Way Elements: The Technology of Metalloids* (Garden City, N.Y.: Doubleday & Company, Inc., 1969), 188 pp., $2.45.

Cotton, Frank A., and Geoffrey Wilkinson, *Advanced Inorganic Chemistry: A Comprehensive Text*, 2d rev. and augm. ed. (New York: Interscience Publishers, John Wiley & Sons, Inc., 1966), $14.50.

Day, M. Clyde, Jr., and Joel Selbin, *Theoretical Inorganic Chemistry*, 2d ed. (New York: Van Nostrand-Reinhold Company, 1969), 590 pp., $12.95.

Dodd, Robert E., and Percy L. Robinson, *Experimental Inorganic Chemistry: A Guide to Laboratory Practice* (New York: American Elsevier Publishing Co., Inc., 1954), $8.25.

Ebsworth, E. A. V., A. G. Maddock, and A. G. Sharpe, *New Pathways in Inorganic Chemistry* (New York: Cambridge University Press, 1968), 390 pp., $13.

Harvey, Kenneth B., and Gerald B. Porter, *Introduction to Physical Inorganic Chemistry* (Reading, Mass.: Addison-Wesley Publishing Co., Inc., 1963), Addison-Wesley series in chemistry, $9.95.

Hill, H. A. O., and P. Day, eds., *Physical Methods in Advanced Inorganic Chemistry* (New York: John Wiley & Sons, Inc., 1968), $16.50.

Holloway, John H., *Noble-Gas Chemistry* (New York: Barnes & Noble, Inc., 1968), 213 pp., $6.75.

Hume-Rothery, William, *Electrons, Atoms, Metals, and Alloys*, 3rd rev. ed. (New York: Dover Publications, Inc., 1963), $3.00 pb.

Hyman, Herbert H., ed., *Noble Gas Compounds* (Chicago: University of Chicago Press, 1963), $12.50.

Johnson, D. A., *Some Thermodynamic Aspects of Inorganic Chemistry* (New York: Cambridge University Press, 1968), 224 pp., $7.50.

Johnson, Ronald C., *Introductory Descriptive Chemistry: "Selected Nonmetals, Their Properties and Behavior"* (Menlo Park, Calif.: W. A. Benjamin, Inc., 1966), The general chemistry monograph series, $4.95; $2.45 pb.

Jolly, William L., *Preparative Inorganic Reactions, Volume 4* (New York: John Wiley & Sons, Inc., 1968), 301 pp., $14.

Jolly, William L., *Preparative Inorganic Reactions, Volume 5* (New York: Interscience Publishers, John Wiley & Sons, Inc., 1968), 248 pp., $24.50.

Kavanau, J. Lee, *Water and Solute-water Interactions* (San Francisco, Holden-Day, Inc., 1964), $5.50.

Krebs, H., and P. H. L. Walter, *Fundamentals of Inorganic Crystal Chemistry* (New York: McGraw-Hill Book Company, Inc., 1968), 405 pp., $13.50.

Larsen, Edwin M., *Transitional Elements* (Menlo Park, Calif.: W. A. Benjamin, Inc., 1965), The general chemistry monograph series, $6.00; $2.95 pb.

Luis, Jose, and Marisa Amoros, *Molecular Crystals: Their Transforms and Diffuse Scattering* (New York: John Wiley & Sons, Inc., 1968), 479 pp., $22.50.

Rich, Ronald, *Periodic Correlations* (Menlo Park, Calif.: W. A. Benjamin, Inc., 1965), The physical inorganic chemistry series, $9.00; $4.95 pb.
Sands, D. E., *Introduction to Crystallography* (Menlo Park, Calif.: W. A. Benjamin, Inc., 1969).
Watt, George W., and William F. Kieffer, *Journal of Chemical Education*, Collected readings in inorganic chemistry (Easton, Pa.: Chemical Education Publishing Company, 1962), $3.50.

GENERAL CHEMISTRY TEXTBOOKS

Anderson, R. T., *Fundamentals of Modern Chemistry* (Glenview, Ill.: Scott, Foresman and Company, 1971).
Bigelow, M. Jerome, *Basic Concepts of Chemistry* (Belmont, Calif.: Bogden & Quigley, Inc., Publishers, 1971).
Brescia, Frank, J. Arents, H. Meislich, et. al., *Fundamentals of Chemistry: A Modern Introduction* (New York: Academic Press, Inc., 1966), $9.95.
Brown, Theodore L., *General Chemistry*, 2d ed. (Columbus, Ohio: Charles E. Merrill, Publishers, 1968), The Merrill physical and inorganic chemistry series), $9.95.
Burman, Harold G., *Principles of General Chemistry* (Boston: Allyn & Bacon, Inc., 1968).
Campbell, Richard D., *College Chemistry—A Survey* (New York: Harcourt Brace Jovanovich, Inc., 1968).
Chemical Bond Approach Project, *Chemical Systems* (New York: McGraw-Hill, Book Company, Inc., 1964), $7.96.
Cherim, Stanley M., *Chemistry for Lab Technicians* (Philadelphia: W. B. Saunders Company, 1971).
Dewar, Michael J. S., *An Introduction to Modern Chemistry* (New York: Oxford University Press, 1965), $2.95 pb.
Dillard, Clyde R., and David E. Goldberg, *Chemistry* (New York: The Macmillan Company, 1971).
Drago, Russell S., *Prerequisites for College Chemistry* (New York: Harcourt Brace Jovanovich, Inc., 1966), $3.50 pb.
Frank, Andrew J., *Qualitative Concepts in General Chemistry* (Columbus, Ohio: Charles E. Merrill Publishers, 1971).
Garard, Ira D., *Invitation to Chemistry* (Garden City, N.Y.: Doubleday & Company, Inc., 1969), 420 pp., $7.95.
Garrett, Alfred B., W. T. Lippincott, and Frank Henry Verhoek, *Chemistry: A Study of Matter* (Waltham, Mass.: Ginn/Blaisdell, 1968).
Hamm, Donald I., *Fundamental Concepts of Chemistry* (New York: Appleton-Century-Crofts, 1969), 811 pp., $9.95.
Horrigan, Phillip A., *Challenge of Chemistry* (New York: McGraw-Hill Book Company, 1970).
Howald, Reed A., and Walter A. Manch, *The Science of Chemistry* (New York: The Macmillan Company, 1971).
Jones, W. Norton, Jr., *Textbook of General Chemistry* (St. Louis, Mo.: The C. V. Mosby Co., 1969), 663 pp., $9.85.
Kask, Uno, *Chemistry: Structure and Changes of Matter* (New York: Barnes & Noble, Inc., 1969).

Keenan, Charles W., and Jesse H. Wood, *General College Chemistry* (New York: Harper & Row, Publishers, 1971).
Kieffer, William F., *Chemistry: A Cultural Approach* (New York: Harper & Row, Publishers, 1971).
Lee, Orden, Ragsdale, *General and Organic Chemistry* (Philadelphia: W. B. Saunders Company, 1971).
Mahan, Bruce H., *University Chemistry*, 2d ed. (Reading, Mass.: Addison-Wesley Publishing Company, Inc., 1969), 832 pp., $11.50.
Maas, Michael L., *Essentials of Chemistry* (Dubuque, Iowa: William C. Brown Company, Publishers, 1971).
Masterson, William L., and Emil J. Slowinski, *Chemical Principles*, 2d ed. (Philadelphia: W. B. Saunders Company, 1969), $8.75.
Miller, Glenn H., *Chemistry* (New York: Harper & Row, Publishers, 1969), 418 pp., $9.95.
Murphy, Daniel, *Foundations of College Chemistry* (New York: The Ronald Press Company, 1969), 676 pp., $10.50.
Nebergall, William H., Frederic C. Schmidt, and Henry F. Holtzclaw, *College Chemistry with Qualitative Analysis*, 4th ed. (Lexington, Mass.: D. C. Heath & Company, 1971).
Quagiano, J. V., and L. M. Vallerino, *Chemistry*, 3rd ed. (Englewood Cliffs, N.J.: Prentice-Hall, Inc., 1969), 844 pp., $11.95.
Paul, Armine D., *General Chemistry* (Columbus, Ohio: Charles E. Merrill Publishers, 1971).
Pauling, Linus, *General Chemistry* (San Francisco: W. H. Freeman & Co., Publishers, 1970).
Pierce, James B., *Chemistry of Matter* (Boston: Houghton Mifflin Company, 1970).
Runquist, Olaf, Clifford J. Creswell, and J. Thomas Head, *Chemical Principles: A Programmed Text, Parts I and II* (Minneapolis, Minn.: Burgess Publishing Co., 1968), 228 pp., $9.75.
Gienko, Michell J., and Robert A. Plane, *Chemistry* (New York: McGraw-Hill Book Company, Inc., 1971).
Spinar, Leo H., *College Chemistry* (Glenview, Ill.: Scott, Foresman and Company, 1968).
Williams, Arthur L., and Hartland D. Embree, *General Chemistry* (Reading, Mass.: Addison-Wesley Publishing Company, Inc., 1970).

LABORATORY MANUALS
Barbour, R., *Glass Blowing for Laboratory Technicians* (Elmsford, N.Y.: Pergamon Press, Inc., 1968), 245 pp., $5.00.
Biechler, Sydney S., *The Behavior of Matter: Lab Experiments in Introductory Chemistry* (Boston: Houghton Mifflin Company, 1969).
Chemical Bond Approach Project, *Investigating Chemical Systems* (New York: McGraw-Hill Book Company, 1959).
Davis, Joseph E., and W. Keith McNab, *Laboratory Manual for Chemistry: Experiments and Principles* (D. C. Heath & Company, 1968).
Drago, Russell S., and Theodore L. Brown, *Experiments in General Chemistry*, rev. ed. (Boston: Allyn & Bacon, Inc., 1965), $5.25 pb.

Flasks, volumetric, 2 (1,000 ml);
  1 (500 ml); 24 (250 ml);
  1 (100 ml)
Funnels, plastic
Funnels, separatory, stoppered
Heaters, electric
Labels, gummed, 12 boxes
Magnets, bar, 12 pairs
Mortars, and pestles, 6 (porcelain,
  6" diam.); 1 (iron, 6" diam.)
pH Meter, 1
Pipets, 24 (10 ml)
Pipet, bulb, 1
Pipet, graduated, 1 (10 ml)
Platinum wire, No. 24, 5 ft.
Retorts, glass stoppered, Pyrex,
  24 (250 ml)
Ring stands, with 3 rings each, 60
Rod, glass, 2 lb. (6mm diam.),
  2 lb. (3mm diam.)
Scissors, 1 pr. (6")
Screwdriver
Shield, metal, for distilling flask
Shields, wire-mesh reinforced
  glass, 2
Spatula, porcelain (25 cm)
Stirrers, electric
Stopper, rubber, assorted sized
  0-6, solid, 1-hole, and 2-hole,
  6 lb.
Suction (outlet) pump, or
  aspirators
Thermometers, 24 ($-20°C$ to
  $110°C$); 2 ($-10°C$ to $360°C$)
Tripods, with concentric rings,
  6 (8")
Troughs, pneumatic, 24
Tubes, drying, calcium chloride,
  24
Tubing, glass, 5 lb. (5mm); 10 lb.
  (6mm)
Tubing, rubber, thin wall, 3/16"
  inside diameter, 10 ft.
Tubing, rubber, medium wall,
  3/16" inside diameter,
  60 ft.
Tubing, rubber, medium wall, ¼"
  inside diameter, 180 ft.
Vials, shell, 30 (3 dram capacity)
Voltmeters, D.C., 0-1.5v
Water still, automatic
Wire gauze, bright, plain, 24

Appendix B

Student Equipment

The following equipment should be available for each student in the laboratory. This equipment may either be stored in a drawer that is moved from a central cabinet or in a drawer provided at the student's work station. The school also has the option of providing each student with a complete set of equipment or having one complete set of equipment for each laboratory station. The former arrangement allows each individual student to be charged for breakage whereas the latter assumes that the school will bear the cost of replacing any broken materials.

Asbestos square (6" x 6" x 1/16")
Beakers 1 (50 ml); 1 (150 ml); 1 (250 ml); 1 (400 ml)
Blowpipe and tip
Bottles, wide mouth, 4 (6 oz)
Brush, test tube
Burner, Bensen and 2 ft tubing
Clamp, Buret
Clamp, test tube
Crucible and cover (size 0)
Dishes, evaporating, 2 (size 0)
File, triangular
Filter paper, 1 pkg. (12.5 cm)
Flasks, Erlenmeyer, 2 (125 ml)
Flask, Florence (250 ml)
Forceps (5")
Funnel (65 mm, short stem)
Funnel tube or thistle tube
Glass plates 4 (3" x 3" x 1/8")
Graduated cylinder, 1 (10 ml); 1 (100 ml)
Magnifying lens
Medicine dropper
Metric ruler (plastic, 15 cm)
Rubber connectors, 2 (thin wall 3/16" inside diam.)
Safety glasses
Spatula, spoon at one end (6")
Splints, wooden, bundle of 50
Spoon, deflagrating
Test tubes, (13 mm x 100 mm) 18 mm x 150 mm); (25 mm x 200 mm)
Test tube rack
Tongs
Triangle, pipestem (2" on a side)
Tubing, deliver, rubber, 2 ft. (medium wall, 3/16" inside diam.)
Watch glasses, 2 (3")
Wing top, or flame spreader
Wire gauze, asbestos center (5" x 5")
Wire triangle

Appendix C

Experiment Requirements Sheets

CONSUMABLE ITEMS

Expt. Title ――――― Expt. No. ___ Date form filled out ___ By: ___	Final Conc.	Amount Needed for One Class	Amount Needed for Entire Week	No. of Grams or Mls. for One Liter	Type of Solution Preparation (A, B, C,)	Shelf Life of Solution

A = Quantitative—Use analytical balance and volumetric glassware
B = Semiquantitative—Use centigram balance and graduated cylinder
C = Special—Use specific directions as listed

Expt. Title ——————— Expt. No. ———————	Student Drawers	Under Sink	Bench Top	Hood	Balance Tables	Stock-room	Student Furnish

SAFETY GUIDE FOR CHEMICALS

Expt. Title Expt. No. _____ Chemical	Degree of Toxicity*	Antidote	Directions for Handling	Other Instructions

*Steere, Norman V., *Handbook of Laboratory Safety* (Cleveland: The Chemical Rubber Company, 1970), pp. 442–551.

Hazardous Mistakes a Student Might Make	Corrective Action to Take

Appendix D

Chemicals Used Frequently in the High School Chemistry Laboratory

The following list of chemicals has been compiled from the chemical lists of several laboratory manuals. Although it is not necessary to have each of these chemicals on hand, the teacher should be aware that these chemicals are frequently used. Ideally, the teacher will compile a list of chemicals (including grade and quantity) that will be needed for the next school year. This list should merely be a summation of the detailed needs list for each experiment. The inventory of chemicals should be kept to a minimum, but should include a sufficient diversity to provide flexibility for the laboratory program.

Acetic acid, glacial
Acetone
Agar-Agar
Albumin, from eggs, soluble, powder
Alizarin, Yellow R
Aluminum, metal, granular
Aluminum chloride
Aluminum nitrate
Aluminum sulfate, powder
Aluminum sulfide
Ammonia-water solution (Ammonium hydroxide), concentrated
Ammonium carbonate, powder
Ammonium chloride, granular
Ammonium dichromate
Ammonium nitrate, reagent
Ammonium oxalate
Ammonium sulfate, granular, white
Ammonium sulfide, light
Ammonium thiocyanate, crystal
Aniline hydrochloride
Antimony, metal, powder
Antimony, trichloride, crystal
Barium chloride, crystal
Barium hydroxide, crystal
Barium nitrate, crystal
Barium sulfate
Benzene
Bromthymol blue
Bromine, N.F.
1-Butanol
2-Butanol
t-Butanol
Butyric acid, 98-100%
Cadmium nitrate
Calcium carbide
Calcium carbonate (marble chips)
Calcium chloride, anhydrous, granular, 4-8 mesh
Calcium fluoride, powder
Calcium hydroxide
Calcium hypochlorite (or bleaching powder)
Calcium metal
Calcium nitrate, crystal
Calcium oxide, lump, purchase yearly

Calcium sulfate, dry, powder
Camphor, synthetic block
Candles, paraffin (assorted sizes)
Carbon disulfide, purified
Carbon tetrachloride, technical
Charcoal, activated
Charcoal, animal (boneblack), powder
Charcoal, wood, lump
Charcoal, wood powder, technical
Charcoal, wood, sticks, for blowpipe work. 3 doz.
Chromium (III) Acetate
Chromium (III) sulfate, granular, reagent
Citric acid, monohydrate
Cobalt (II) chloride, crystal
Cobalt (II) nitrate, crystal
Copper, metal, foil, 1/100" thick
Copper, metal turnings
Copper, metal, shot
Copper, wire, bare, No. 16, No. 18, No. 22, and/or No. 26
Copper (II) nitrate, crystal
Copper (II) oxide, black, powder
Copper (II) sulfate, anhydrous
Copper (II) sulfate, granular
Cottonseed oil
Cyclohexane
Diarsenic trioxide, powder, N.F.
Dimethylglyoxime
para-Dichlorobenzene
Ethyl acetate
Ethyl alcohol (ethanol), denatured
Flour
Formaldehyde
Formic acid, 85–90%
Gasoline
Gelatin, granular
Glycerol (glycerin), white
Graphite, powder
Hydrochloric acid, concentrated
Hydrogen peroxide, 30%, purchase each year
Hydrogen peroxide, 3%
Indigo carmine
Ink, India, 1 bottle
Iodine, resublimed
Iron, metal, filings, fine
Iron, metal, wire
Iron (II) ammonium sulfate, granular
Iron (II) (III) oxide (magnetite), powdered
Iron (II) sulfate, crystal
Iron (II) sulfide, broken lumps
Iron (III) ammonium citrate, green
Iron (III) chloride, lump
Iron (III) nitrate, crystal
iso-Amyl alcohol, 85%
Kerosene
Lampblack
Lead, metal, foil, 8/1000" thick
Lead, metal, shot
Lead (II) acetate, granular, N.F.
Lead (II) nitrate, crystal
Lead (II) (III) oxide
Lead sulfide, powdered
Lithium, metal
Lithium, nitrate
Litmus paper, blue
Litmus paper, red
Litmus, powder
Magnesium, metal, ribbon
Magnesium chloride, crystal
Magnesium hydroxide, "milk of magnesia," 1 bottle
Magnesium nitrate
Magnesium oxide
Magnesium sulfate, crystal
Malachite green
Manganese dioxide, granular, reagent grade only
Manganese nitrate, crystal
Manganese (II) sulfate

Mercury (I) nitrate, crystal
Mercury (II) chloride, granular
Mercury (II) nitrate, crystal
Methyl alcohol (methanol), absolute, synthetic
Methylene blue
Methylethyl ketone
Methyl orange (indicator)
Methyl red
Molasses
Nickel (II) chloride
Nickel (II) sulfate, crystal
Nitric acid, concentrated
Oxalic acid
Oxygen
Paraffin
Perchloric acid
Phenolphthalein, powder
Phosphorus, red, amorphous
Phosphorus, white (yellow), thick sticks
Phosphorous pentoxide
Phthalic anhydride
Potassium, metal
Potassium bromide, granular
Potassium carbonate, granular
Potassium chlorate, small crystal
Potassium chloride, white, granular
Potassium chromate
Potassium cyanide
Potassium dichromate
Potassium ferricyanide
Potassium ferrocyanide
Potassium hexacyanoferrate (II), crystal
Potassium hexacyanoferrate (III), crystal
Potassium hydrogen carbonate, crystal
Potassium hydrogen oxalate
Potassium hydrogen sulfate
Potassium hydrogen tartrate, powder

Potassium hydroxide, pellets
Potassium iodate
Potassium iodide, crystal
Potassium nitrate, crystal
Potassium permanganate, crystal
Potassium sulfate, crystal
Potassium thiocyanate
Pyrogallol (pyrolallic acid), crystal
Salicylic acid, crystal
Sand
Sandpaper, fine
Silver, metal, foil, 1/100" thick
Silver acetate
Silver nitrate, crystal
Soap
Soda lime
Sodium, metal
Sodium acetate, crystal
Sodium aluminum sulfate
Sodium borate, tetra-, powder
Sodium carbonate, anhydrous
Sodium carbonate, crystal
Sodium chloride, crystal
Sodium chloride, white, technical (table salt)
Sodium chromate
Sodium dichromate
Sodium fluoride
Sodium hydrogen carbonate, powder
Sodium hydroxide, pellets
Sodium hydroxide, flake
Sodium hypochlorite (Chlorox)
Sodium iodide
Sodium metabisulfite
Sodium nitrate, crystal
Sodium nitrite, granular
Sodium oxalate, reagent
Sodium peroxide, reagent, 20 mesh and finer
Sodium phosphate, tribasic
Sodium sulfate, small crystal
Sodium sulfide, crystal

Sodium sulfite, dry, powder
Sodium thiocyanite
Sodium thiosulfate, crystal
Starch, corn
Strontium chloride
Strontium nitrate
Sucrose, granular (granulated sugar)
Sugar, dark brown
Sulfur, sublimed (sulfur flowers)
Sulfur, roll
Sulfuric acid, concentrated
Taconite, powdered ore
Tartaric acid, powder
Thermite, black
Thermite igniting mixture
Thread, cotton, No. 40
Tin, metal, mossy
Tin (II) chloride, crystal
Universal indicator (and color chart)
Uranyl nitrate
Uranium ore
Vinegar
Zeolite water softener
Zinc, metal, mossy
Zinc, metal powder (dust)
Zinc, metal, sheet, 1/100" thick
Zinc chloride, dry, granular
Zinc nitrate, crystal
Zinc sulfate, granular
Zinc sulfide ore, powdered

The major acids; hydrochloric, nitric, and sulfuric should be purchased in plastic capped 2.2 liter bottles. Ammonium hydroxide should also be purchased this way. Each of these substances should be reagent grade to provide the accuracy required in some experiments. The teacher is advised against ordering acids in larger containers as problems may arise in handling such quantities.

Appendix E

Inventory Card for Chemicals and Equipment

Item	
Classification	Number on Hand
Description	
Supplier	Order No.
Remarks	

Appendix F

Solution Preparation Formulas

The following list of solutions is not to be considered as a complete list of solutions that will have to be prepared. Instead, it represents some of the more common solutions that will be required in various experiments. The quantities indicated of each solute will give a solution that is 0.1M unless otherwise indicated. To determine the quantities of solute necessary to prepare other concentrations, the teacher need only multiply by the constant that will give him the desired concentration.

Acid	Concentration	Milliliters of acid/liter
HCl	12M concentrated	1,000.0
HCl	6M	500.0
HCl	5M	416.7
HCl	4M	333.3
HCl	3M	250.0
HCl	2M	166.6
HCl	1M	125.0
HCl	0.9M	75.0
HCl	0.8M	66.7
HCl	0.7M	58.3
HCl	0.6M	50.0
HCl	0.5M	41.7
HCl	0.4M	33.3
HCl	0.3M	25.0
HCl	0.2M	16.7
HCl	0.1M	12.5
H_2SO_4	18M concentrated	1,000.0
H_2SO_4	6M	333.3
H_2SO_4	5M	227.8
H_2SO_4	4M	222.2
H_2SO_4	3M	166.7
H_2SO_4	2M	111.1
H_2SO_4	1M	55.6

Acid	Concentration	Milliliters of acid/liter
H_2SO_4	0.9M	50.0
H_2SO_4	0.8M	44.4
H_2SO_4	0.7M	38.9
H_2SO_4	0.6M	33.3
H_2SO_4	0.5M	27.8
H_2SO_4	0.4M	22.2
H_2SO_4	0.3M	16.7
H_2SO_4	0.2M	11.1
H_2SO_4	0.1M	5.6
HNO_3	16M concentrated	1,000.0
HNO_3	6M	375.0
HNO_3	5M	312.5
HNO_3	4M	250.0
HNO_3	3M	187.5
HNO_3	2M	125.0
HNO_3	1M	62.5
HNO_3	0.9M	56.2
HNO_3	0.8M	50.0
HNO_3	0.7M	43.8
HNO_3	0.6M	37.5
HNO_3	0.5M	31.2
HNO_3	0.4M	25.0
HNO_3	0.3M	18.8
HNO_3	0.2M	12.5
HNO_3	0.1M	6.2
$HC_2H_3O_2$	17M concentrated	1,000.0
$HC_2H_3O_2$	6M	352.9
$HC_2H_3O_2$	5M	294.1
$HC_2H_3O_2$	4M	235.3
$HC_2H_3O_2$	3M	176.4
$HC_2H_3O_2$	2M	117.6
$HC_2H_3O_2$	1M	58.8
$HC_2H_3O_2$	0.9M	52.9
$HC_2H_3O_2$	0.8M	47.1
$HC_2H_3O_2$	0.7M	41.2
$HC_2H_3O_2$	0.6M	35.3
$HC_2H_3O_2$	0.5M	29.4
$HC_2H_3O_2$	0.4M	23.5

Acid	Concentration	Milliliters of acid/liter
$HC_2H_3O_2$	0.3M	17.6
$HC_2H_3O_2$	0.2M	11.7
$HC_2H_3O_2$	0.1M	5.9

Base	Concentration	Milliliters of base/liter
NH_4OH	15M concentrated	1,000.0
NH_4OH	6M	400.0
NH_4OH	5M	333.3
NH_4OH	4M	266.7
NH_4OH	3M	200.0
NH_4OH	2M	133.4
NH_4OH	1M	66.7
NH_4OH	0.9M	60.0
NH_4OH	0.8M	53.3
NH_4OH	0.7M	46.7
NH_4OH	0.6M	40.0
NH_4OH	0.5M	33.3
NH_4OH	0.4M	26.7
NH_4OH	0.3M	20.0
NH_4OH	0.2M	13.3
NH_4OH	0.1M	6.7

Base	Concentration	Grams of base/liter
NaOH	6M	240.0
NaOH	5M	200.0
NaOH	4M	160.0
NaOH	3M	120.0
NaOH	2M	80.0
NaOH	1M	40.0
NaOH	0.9M	36.0
NaOH	0.8M	32.0
NaOH	0.7M	28.0
NaOH	0.6M	24.0
NaOH	0.5M	20.0
NaOH	0.4M	16.0
NaOH	0.3M	12.0
NaOH	0.2M	8.0
NaOH	0.1M	4.0

PREPARATION OF 0.1M SOLUTIONS (1,000 ml)
(Solutions marked with * are not 0.1M)

Aluminum Chloride - $AlCl_3$ - 132 g/mole - 13.2 g of $AlCl_3$

Aluminum Sulfate - $Al_2(SO_4)_3$ - 442 g/mole - 67 g of $Al_2(SO_4) \cdot 18H_2O$

Ammonium Sulfide - $(NH_4)_2S$ - 68 g/mole - 40 g of 8% $(NH_4)_2S$ solution.

Antimony trichloride - $SbCl_3$ - 227.5 g/mole - dissolve 22.8 g of $SbCl_3$ in water containing 25 ml of concentrated HCl. Dilute to 1 liter.

Barium chloride - $BaCl_2$ - 207 g/mole - 24.3 g of $BaCl_2 \cdot 2H_2O$.

Barium hydroxide - $Ba(OH)_2$ - 171 g/mole - 31.5 g of $Ba(OH)_2 \cdot 8H_2O$.

Barium Nitrate - $Ba(NO_3)_2$ - 261 g/mole - 26.1 g of $Ba(NO_3)_2$.

*Bromine water - Br_2 - 160 g/mole - dissolve 1 ml (liquid bromine) in 1 liter of water. Store in a brown bottle.

*Bromthymol blue - 0.04%, Place 0.1 g along with 16 ml of 0.01M NaOH in a mortar and mix. Dilute to 250 ml with water.

Cadmium nitrate - $Cd(NO_3)_2$ - 236 g/mole - 30.8 g of $Cd(NO_3)_2 \cdot 4H_2O$.

Calcium chloride - $CaCl_2$ - 111 g/mole - 22 g of $CaCl_2 \cdot 6H_2O$.

Calcium nitrate - $Ca(NO_3)_2$ - 164 g/mole - 16.4 g of $Ca(NO_3)_2$.

*Calcium sulfate - $Ca(SO_4)$ saturated at 5 g of $CaSO_4 \cdot 2H_2O$ per liter.

*Chlorine water - Cl_2. Procure from commercial source.

Chromium sulfate - $Cr_2(SO_4)_3$ - 392 g/mole - 72 g of $Cr_2(SO_4)_3 \cdot 18H_2O$.

*Cleaning Solution------Add 400 ml of Concentrated sulfuric acid to 50 g of $K_2Cr_2O_7$. Let the glassware soak overnight and then save the solution for reuse. Handle all glassware that has been soaked in cleaning solution with extreme care until the final rinse.

Cobalt (II) Chloride - $CoCl_2$ - 129 g/mole - 24 g of $CoCl_2 \cdot 6H_2O$.

Cobalt (II) nitrate - $Co(NO_3)_2$ - 183 g/mole - 29 g of $Co(NO_3)_2 \cdot 6H_2O$.

Copper (II) nitrate - $Cu(NO_3)_2$ - 187.5 g/mole - 30 g of $Cu(NO_3)_2 \cdot 6H_2O$.

Copper (II) sulfate - $CuSO_4$ - 159.5 g/mole - dissolve 25 g of $CuSO_4 \cdot 5H_2O$ in water containing 0.5 ml of concentrated sulfuric acid. Dilute to 1 liter.

Ethanol - C_2H_5OH - 46 g/mole - 4.8 g of 95% C_2H_5OH and dilute to 1 liter.

Formalin - 40% formaldehyde. 400 g of formaldehyde in 600 g of H_2O.

Gylcerol - $C_3H_5(OH)_3$ - 92 g/mole - 9.2 g of $C_3H_5(OH)_3$ dilute 1 liter.

*Hydrogen peroxide - H_2O_2 - 6% 200 g of 30% H_2O_2 in 800 ml of H_2O.
Hydrogen sulfide (very dilute) - H_2S - 34 g/mole - Use 25 ml of a saturated solution of hydrogen sulfide in water and dilute to one liter.
*Iodine - I_2 - 254 g/mole - 3 g of Kl and .6g of I_2 in 200 ml of H_2O.
Iron (II) ammonium sulfate - $Fe(NH_4)_2$ - 284 g/mole - dissolve 39 g of $Fe(NH_4)_2 (SO_4)_2 \cdot 6H_2O$ in water containing 10 ml of concentrated sulfuric acid. Dilute to one liter (should be freshly prepared).
Iron (II) sulfate - $FeSO_4$ - 152 g/mole - dissolve 28 g of $FeSO_4 \cdot 7H_2O$ in water containing 3 ml of concentrated sulfuric acid. Dilute to 1 liter. Add 3 small iron nails to the solution (should be freshly prepared).
Iron (II) chloride - $FeCl_2$ - 126 g/mole - React steel wool with 6 M HCl permitting excess of steel wool to remain to keep all iron ions in Fe(II) state.
Iron (III) chloride - 161.5 g/mole - $FeCl_3$ - Dissolve 27 g of $FeCl_3 \cdot 6H_2O$ in water containing 20 ml of concentrated HCl. Dilute to 1 liter.
Lead (II) acetate - $Pb(C_2H_3O_2)_2$ - 266 g/mole - 38 g of $Pb(C_2H_3O_2)_2 \cdot 3H_2O$.
Lead (II) nitrate - $Pb(NO_3)_2$ - 331 g/mole - 33.1 g of $Pb(NO_3)_2$.
*Limewater - Add 5 grams of $Ca(OH)_2$ to 1 liter of water. Shake well and allow the undissolved solids to settle. Filter the saturated solution into another labeled bottle.
Lithium nitrate - $LiNO_3$ - 69 g/mole - 6.9 g of $LiNO_3$.
Magnesium Chloride - $MgCl_2$ - 94 g/mole - 20 g of $MgCl_2 \cdot 6H_2O$.
*Magnesium hydroxide - (saturated), silute commercial Milk of Magnesia and filter.
Magnesium sulfate - $MgSO_4$ - 120 g/mole - 24.6 g of $MgSO_4 \cdot 7H_2O$.
Manganese (II) sulfate - $MnSO_4$ - 151 g/mole - 27.6 g of $MnSO_4 \cdot 7H_2O$.
Mercury (I) nitrate - $HgNO_3$ - 263 g/mole - dissolve 28.1 g of $HgNO_3 \cdot H_2O$ in water to which 50 ml of concentrated nitric acid has been added. Dilute to 1 liter.
*Methyl orange - 1 g of methyl orange in 1 liter of H_2O.
*Oxalic acid - (saturated) $H_2C_2O_4$ - 90 g/mole - 100 g of $(COOH)_2 \cdot 2H_2O$.
*Phenolphthalein - Dissolve 10 g of phenolphthalein in 500 ml of denatured alcohol and dilute with 500 ml of water.
Potassium bromide - KBr - 199 g/mole - 11.9 g of KBr.
Potassium carbonate - K_2CO_3 - 138 g/mole - 13.8 g of K_2CO_3.

Potassium chloride - KCl - 74 g/mole - 7.4 g of KCl.
Potassium chromate K_2CrO_4 - 194 g/mole - 19.4 g of K_2CrO_4.
Potassium cyanide KCN - 65 g/mole - 6.5 g of KCN.
Potassium hexacyanoferrate (II) - $K_4Fe(CN)_6$ - 368 g/mole - 42 g of $K_4Fe(CN)_6 \cdot 3H_2O$.
Potassium hexacyanoferrate (III) - $K_3Fe(CN)_6$ - 329 g/mole - 39 g of $K_4Fe(CN)_6 \cdot 3H_2O$.
Potassium hydrogen tartrate - $KHC_4H_4O_6$ - 188 g/mole - 18.8 g of $KHC_4H_4O_6$.
Potassium hydroxide - KOH - 56 g/mole - 5.6 g of KOH.
Potassium iodide - Kl - 166 g/mole - 16.6 g of Kl.
Potassium nitrate - KNO_3 - 101 g/mole - 10.1 g of KNO_3.
Potassium permanganate - $KMnO_4$ - 158 g/mole - 15.8 g of $KMnO_4$.
*Silver acetate - $AgC_2H_3O_2$ - 167 g/mole - (saturated) 11 g of $AgC_2H_3O_2$.
Silver nitrate - $AgNO_3$ - 170 g/mole - 17 g of $AgNO_3$. Store in a brown bottle.
Sodium acetate - $NaC_2H_3O_2$ - 82 g/mole - 8.2 g of fused $NaC_2H_3O_2$.
Sodium aluminum sulfate - $NaAl(SO_4)_2$ - 242 g/mole - 46 g of $NaAl(SO_4)_2 \cdot 12H_2O$.
Sodium carbonate - Na_2CO_3 - 106 g/mole - 10.6 g of Na_2CO_3.
Sodium chloride - NaCl - 58 g/mole - 5.8 g of NaCl.
Sodium fluoride - NaF - 42 g/mole - 4.2 g of NaF.
Sodium hydrogen carbonate - $NaHCO_3$ - 84 g/mole - 8.4 g of $NaHCO_3$.
Sodium nitrate - $NaNO_3$ - 85 g/mole - 8.5 g of $NaNO_3$.
Sodium oxalate - $Na_2C_2O_4$ - 134 g/mole - 13.4 g of $Na_2C_2O_4$.
Sodium phosphate - Na_3PO_4 - 164 g/mole - 37.5 g of $Na_3PO_4 \cdot 12H_2O$.
Sodium sulfate Na_2SO_4 - 142 g/mole - 14.2 g of Na_2SO_4.
Sodium sulfide - Na_2S - 78 g/mole - 24 g of $Na_2S \cdot 9H_2O$.
Sodium tetraborate - $Na_2B_4O_7$ - 202 g/mole - 38.2 g of $Na_2B_4O_7 \cdot 10H_2O$.
Sodium thiosulfate - $Na_2S_2O_3$ - 158 g/mole - 25 g of $Na_2S_2O_3 \cdot 5H_2O$.
*Starch - Place 0.01 grams of HgI_2 and 2 grams of soluble starch in a small amount of water and stir into a paste. The mixture should be added slowly to 1 liter of boiling water and boiled for a few minutes. Store in a glass-stoppered bottle. Prepare fresh weekly.
Strontium chloride - $SrCl_2$ - 158 g/mole - 26.7 g of $SrCl_2 \cdot 6H_2O$.
Strontium nitrate - $Sr(NO_3)_2$ - 212 g/mole - 21.2 g of $Sr(NO_3)_2$.
Sucrose (Sugar) - $C_{12}H_{22}O_{11}$ - 342 g/mole - 34.2 g of $C_{12}H_{22}O_{11}$.
Tartaric Acid - $H_2C_4H_4O_6$ - 150 g/mole - 15 g of $H_2C_4H_4O_6$.

Tin (II) Chloride - $SnCl_2$ - 189 g/mole - Dissolve 23 g of $SnCl_2 \cdot 2H_2O$ in 35 ml of concentrated hydrochloric acid. Dilute to 1 liter with water. Add a few granules of tin (should be freshly prepared).
Zinc chloride - $ZnCl_2$ - 135 g/mole - 14 g of $ZnCl_2$.
Zinc nitrate - $Zn(NO_3)_2$ - 189 g/mole - 30 g of $Zn(NO_3)_2 \cdot 6H_2O$.
Zinc sulfate - $ZnSO_4$ - 161 g/mole - 29 g of $ZnSO_4 \cdot 7H_2O$.

Appendix G

Student Safety Rules for the Chemistry Laboratory
1. Always wear safety glasses when handling chemicals. Wear aprons or gloves as the instructor advises.
2. Keep your appearance as neat as possible. Wear snug-fitting clothing and keep long hair tied behind the head.
3. Flush all chemicals off your skin and clothing immediately with plenty of water.
4. Do not point test tubes at your neighbors while heating liquids of any kind.
5. Do not hold your face over an open container. All fumes and odors should be observed cautiously.
6. Use only the prescribed quantities of chemicals. Never attempt unauthorized experiments.
7. Insert glass tubing and thermometers into rubber stoppers with great caution. Follow the directions of the instructor.
8. Dilute acids by adding the concentrated acid to distilled water.
9. Never return unused chemicals to the original containers.
10. Dispose of extra and waste chemicals as directed.
11. Do not taste any chemicals in the laboratory.
12. Be cautious with heated glass; it may not be as cool as it looks.
13. Use the hood for experiments that produce noxious gases.
14. Restrict your conversations to the experiment. Socializing should be conducted after class.
15. Know the location of all safety equipment.

Appendix H

Accident Record Form

Date:	Time:
Name:	
Nature of injury:	
Treatment given:	
Follow up:	
Cause of accident:	
Nature of experiment being conducted:	

Appendix I

Checklist for Shutting Down the Laboratory

1. Return all pure chemicals to the stockroom.
2. Dispose of extra solutions that are not needed.
3. Wash and reshelve all glassware.
4. Remove all reagents from the laboratory to the stockroom.
5. Clean all student glassware and replace missing components.
6. Clean bench tops and clear display areas.
7. Compile the semester or yearly usage lists.
8. Remove all waste chemicals and solutions from the laboratory's preparation room and stockroom.
9. Check the labels on all bottles for tightness and legibility.
10. Inspect all bottles for cracks and/or leakage.
11. Lock the chemical stockroom door.
12. Shut off the electricity, gas, and water for the student desks and the preparation area.
13. Inform the administration and custodial personnel where chemicals and equipment received during the summer should be stored.

Appendix J

Sources of Chemistry and Physics Equipment and Supplies

The following lists of sources of equipment and supplies are separated to enable the teacher to be sure that he is dealing with a company that deals primarily in the given area. Many of the companies in these lists can supply the needs of both the chemistry and physics teacher. Up-to-date catalogues should be requested from companies supplying the desired equipment and supplies.

NAMES AND ADDRESSES OF SUPPLIERS OF CHEMICAL EQUIPMENT AND SUPPLIES

Ainsworth & Sons
2151 Lawrence Street
Denver, Colorado 80205
(equipment)

Allied Chemical & Dye Corp.
Box 111
Ashland, Kentucky 41101
(chemicals)

Cambosco Scientific Company
342 Western Avenue
Brighton, Massachusetts
(equipment)

Central Scientific Company
2600 S. Kostner Avenue
Chicago, Illinois 60623
(Chemistry & Physics supplies)

Chemical Rubber Company
2310 Superior Avenue
Cleveland, Ohio 44114
(chemistry laboratory equipment)

Clay-Adams Company
141 E. 25th Street
New York, New York 10010
(equipment)

Cole-Larmer Instrument and Equipment Co.
7330 North Clark Street
Chicago, Illinois 60626
(equipment)

Corning Glass Works
Corning, New York 14830
(glass equipment)

Denoyer-Gerppert Co.
5235 N. Ravenswood Avenue
Chicago, Illinois 60640
(globes and charts)

Difco Laboratories, Incorporated
920 Henry Street
Detroit, Michigan 48201
(supplies)

Dow Chemical Company
Midland, Michigan 48232
(chemical supplies)

Ealing Corporation
2225 Massachusetts Avenue
Cambridge, Massachusetts 02140

Eastman Kodak Company
Motion Picture and Education
  Markets Division
Rochester, New York 14650
(chemicals)

Edmund Scientific Company
Barrington, New Jersey 08003
(equipment)

Fisher Scientific Supply
  Company
711 Forbes Avenue
Pittsburgh, Pennsylvania 15219
(laboratory equipment and
  supplies)

Graf-Apsco Company
5868 N. Broadway
Chicago, Illinois 60640
(equipment)

Harshaw Chemical Company
6265 Wiche Road
Cincinnati, Ohio 45237
(supplies)

Kaud K. Laboratories, Inc.
121 Express Street
Engineers Hill
Plainview, New York 11803
(rare chemicals and serums)

Kaufman Glass Company
1209-21 French Street
Wilmington, Delaware
(plastic and glass laboratory
  materials)

Klinger Scientific Apparatus
83-45 Parsons Boulevard
Jamaica, New York 11432
(imported science apparatus)

Lab-Lone Instruments
15th and Bloomingdale
Melrose Park, Illinois 60160
(instruments)

Lab Industries
1802 2nd Street
Berkeley, California 94710
(laboratory equipment)

Lab Com Company, Incorporated
Box 14466
Houston, Texas 77021

Laboratory Equipment
Box 1511
St. Joseph, Michigan 49085

Laboratory Metalware
  Corporation
429 W. Superior Street
Chicago, Illinois 60610

Lab Research Manufacturing
Northwest Boulevard
Vineland, New Jersey 08360
(chemicals)

Lab Sciences, Incorporated
604 Park Drive
Boca Raton, Florida 33432
(equipment)

Charles Lane Corporation
105 Chambers Street
New York, New York

Lederle Laboratories,
Division American Cyanamid
  Company
North Middletown Road
Pearl River, New York 10965
(biological supplies)

Macalester Scientific Corporation
60 Arsenal Street
Watertown, Massachusetts 02072
(general chemistry and physics
  equipment)

Merck and Company,
   Incorporated
Lincoln Avenue
Rahway, New Jersey 07065
(chemical supplies)

Monsanto Chemical Company
1700 S. 2nd Street
St. Louis, Missouri 63104
(chemical supplies)

New York Scientific Supply
   Company
331 East 38th Street
New York, New York
(general science supplies)

Scientific Kit Company
402 Greentree Road
Pittsburgh, Pennsylvania 15220
(general laboratory equipment
   and supplies)

Polaroid Corporation
730 Main Street
Cambridge, Massachusetts 02139
(film, cameras)

Kimble Glass Company
Owen-Illinois, Incorporated
P. O. Box 1035
Toledo, Ohio
(glass)

LaPine Scientific Company
6001 S. Knox Avenue
Chicago, Illinois 60629
(chemistry, biology, and physics
   equipment and supplies)

Ohaus Scale Corporation
1050 Commerce Avenue
Union, New Jersey
(scales and balances)

Pfizer, Charles & Company
11 Bartlett Street
Brooklyn, New York 11206
(chemical supplies)

Sargent, E. H. and Company
4647 West Foster Avenue
Chicago, Illinois 60630
(chemistry laboratory
   equipment)

Science Electronics,
   Division of General Electronic
   Laboratories, Incorporated
1085 Commonwealth Avenue
Boston, Massachusetts 02215
(physics and apparatus)

Science Kit, Incorporated
Box 69
Tonawanda, New York
(general elementary and junior
   high science equipment)

Scientific Kit Company
402 Greentree Road
Pittsburgh, Pennsylvania 15220
(general laboratory equipment
   and supplies)

Standard Scientific Corporation
34 W. 4th Street
New York, New York
(laboratory supplies)

Southwest Scientific Company
319 E. Indian Plaza
Scottsdale, Arizona 85251
(minerology specimens and
   supplies)

Teaching Materials Corporation
575 Lexington Avenue
New York, New York
(equipment)

Thomas, Arthur H., Company
Box 779
Philadelphia, Pennsylvania
   19105
(general laboratory supplies and
   equipment)

Torsion Balance Company
Clifton, New Jersey
   07013
(balances and general physics
   equipment)

Will Scientific, Incorporated
Box 1050
Rochester, New York
(laboratory equipment and
   supplies)

NAMES AND ADDRESSES OF SUPPLIERS OF
PHYSICS EQUIPMENT AND MATERIALS

Central Scientific Company
2600 S. Kostner Avenue
Chicago, Illinois 60623

Damon Engineering, Inc.
   Educational Division
115 Fourth Avenue
Needham, Massachusetts 02194

Ealing Corporation
2225 Massachusetts Avenue
Cambridge, Massachusetts 02130

Edmund Scientific Company
Barrington, New Jersey 08007

Eduquip/Linco
1220 Adams Street
Boston, Massachusetts 02124

Heath Company
Benton Harbor, Michigan 49022

Holt, Rinehart and Winston, Inc.
383 Madison Avenue
New York, New York 10017

H-P Scientific Company
74 Penarrow Drive
Tonowanda, New York 14150

Klinger Scientific Apparatus
   Corporation
83-45 Parsons Boulevard
Jamaica, New York 11432

La Pine
6001 S. Knox Avenue
Chicago, Illinois 60629

Macalester Scientific Company
Route 111 and Everett Turnpike
Nashua, New Hampshire 03060

PASCO Scientific
1933 Republic Avenue
San Leandro, California 94577

Stansi Scientific Division
1231 North Honore Street
Chicago, Illinois 60622

Thornton Associates, Inc.
87 Beaver Street
Waltham, Massachusetts 02154

Welch Scientific Company
7300 N. Lindner Avenue
Skokie, Illinois 60076

Appendix K

Directory of Educational Publishers

The following companies supply most of the books used in high school chemistry and physics courses. Some of these companies supply paperback books that may be used to supplement the chosen text. Other companies deal primarily in college level science texts that may be used by the teacher as a resource tool.

Academic Press Inc.
111 Fifth Avenue
New York, New York 10003

Addison-Wesley Publishing
  Co., Inc.
Reading, Massachusetts 01867

Allyn & Bacon, Inc.
400 Atlantic Avenue
Boston, Massachusetts 02210

Appleton-Century-Crofts
440 Park Avenue South
New York, New York 10016

Bantam Books, Inc.
666 Fifth Avenue
New York, New York 10019

Barnes and Noble
105 Fifth Avenue
New York, New York 10003

Benefic Press
10300 W. Rooseveldt Road
Westchester, Illinois 60153

W. A. Benjamin, Inc.
2 Park Avenue
New York, New York 10016

Chas. A. Bennett Company
809 W. Detweiller Drive
Peoria, Illinois 61614

The Bobbs-Merrill Co., Inc.
4300 W. 62nd Street
Indianapolis, Indiana 46268

The Bruce Publishing Company
850 Third Avenue
New York, New York 10022

Burgess Publishing Co.
426 South 6th Street
Minneapolis, Minnesota 55415

Chemical Publishing Co., Inc.
200 Park Avenue South
New York, New York 10003

Chemical Rubber Co.
18901 Cranwood Parkway
Cleveland, Ohio 44128

College Entrance Book Co.
104 Fifth Avenue
New York, New York 10011

The George F. Cram Co., Inc.
301 South LaSalle Street
Indianapolis, Indiana 46206

Denoyer-Geppert Co.
5235 Ravenswood Avenue
Chicago, Illinois 60640

Doubleday and Co., Inc.
501 Franklin Avenue
Garden City, New York 11530

The Economy Company
1901 N. Walnut Avenue
Oklahoma City, Oklahoma 73105

Encyclopedia Britannica Educational Corporation
425 N. Michigan Avenue
Chicago, Illinois 60611

Fawcett Publications, Inc.
67 W. 44th Street
New York, New York 10036

Fearon Publishers, Inc.
2165 Park Boulevard
Palo Alto, California 94306

Follett Educational Corporation
1010 W. Washington Boulevard
Chicago, Illinois 60607

Ginn and Company
Statler Bldg.
Back Bay P. O. 191
Boston, Massachusetts 02117

Globe Book Company, Inc.
175 Fifth Avenue
New York, New York 10010

Gregg Division—McGraw-Hill Book Co.
Manchester Road
Manchester, Missouri 63011

Grune and Stratton, Inc.
111 Fifth Avenue
New York, New York 10003

Grolier Educational Corporation
575 Lexington Avenue
New York, New York 10022

E. M. Hale and Company
1201 S. Hastings Way
Eau Claire, Wisconsin 54701

Hammond Incorporated
Maplewood, New Jersey 07040

Harcourt Brace Jovanovich, Inc.
757 Third Avenue
New York, New York 10017

Harper & Row Publishers
10 East 53rd Street
New York, New York 10017

D. C. Heath and Co.
Division Raytheon Education Company
125 Spring Street
Lexington, Massachusetts 02173

Holt, Rinehart and Winston, Inc.
383 Madison Avenue
New York, New York 10017

Houghton Mifflin Co.
2 Park Street
Boston, Massachusetts

Instructor Publications, Inc.
Bank Street
Danville, New York 14437

Laidlaw Brothers
Division of Doubleday & Company, Inc.
Thatcher and Madison Streets
River Forest, Illinois 60305

J. B. Lippincott Co.
E. Washington Square Street
Philadelphia, Pennsylvania 19105

Little, Brown and Co., Inc.
34 Beacon Street
Boston, Massachusetts 02106

Lyons and Carnahan
407 E. 25th Street
Chicago, Illinois 60616

Macmillan Publishing Co., Inc.
866 Third Avenue
New York, New York 10022

McCormick-Mathers Publishing
    Company
300 Pike Street
Cincinnati, Ohio 45202

McGraw-Hill Book Company,
    Inc.
1221 Avenue of Americas
New York, New York

Charles E. Merrill Books, Inc.
1300 Alum Creek Drive
Columbus, Ohio 43216

National Education Association
1201 16th Street, N. W.
Washington, D. C. 20036

A. J. Nystrom & Co.
3333 Elston Avenue
Chicago, Illinois 60618

Oxford Book Co., Inc.
Keystone Education Press Inc.
387 Park Avenue South
New York, New York 10016

Popular Science Publishing
    Co., Inc.
A. V. Division
355 Lexington Avenue
New York, New York 10017

Prentice-Hall, Inc.
Englewood Cliffs,
    New Jersey 07632

Rand McNally and Company
P. O. Box 7600
Chicago, Illinois 60680

Random House School &
    Library Service, Inc.
201 East 50th Street
New York, New York 10022

William H. Sadlier, Inc.
300 West Washington Street
Chicago, Illinois 60606

W. B. Saunders Company
W. Washington Square
Philadelphia, Pennsylvania
    19105

Scholastic Book Services
50 W. 44th Street
New York, New York 10036

Science Research Associates,
    Inc.
259 E. Erie Street
Chicago, Illinois 60611

Scott Foresman and Company
1900 E. Lake Avenue
Glenview, Illinois 60025

Silver Burdett Company
250 James Street
Morristown, New Jersey
    07960

The L. W. Singer Co., Inc.
201 E. 50th Street
New York, New York 10022

South-western Publishing Co.
500 W. Harrison Street
Chicago, Illinois 60644

Time-Life Books,
A Division of Time Inc.
Time & Life Bldg.
Rockefeller Center
New York, New York 10020

Van Nostrand Reinhold
    Company
450 W. 33rd Street
New York, New York 10001

John Wiley & Sons, Inc.
605 Third Avenue
New York, New York 10016

Xerox Education Division
1200 High Ridge Road
Stamford, Connecticut

Appendix L

Sources of Motion Pictures and Slidefilms—Free Loan

The following companies have indicated that they will supply films, free of charge, to teachers for classroom showing. Many of these films are of a specialized nature and contain a significant amount of advertising. This list is not complete but should provide many sources for the teacher. Not all of these films are directly applicable to chemistry or physics classes, but they could be of use in other classes the teacher may be called upon to teach. Brochures should be obtained from the companies supplying films that could be of interest.

Admiral Corporation
3800 Corlandt Street
Chicago, Illinois 60647

Aetna Life & Casualty
151 Farmington Avenue
Hartford, Connecticut 06115

Air Reduction
2001 W. 16th Street
Broadview, Illinois 60153

Alaska Railroad
Department of Interior
Room 4140
Washington, D.C. 20240

Allis-Chalmers Mfg. Company
Tractor Group
Milwaukee, Wisconsin 53201

Allstate Insurance Company
Allstate Plaza
Northbrook, Illinois 60062

Alpex Corporation
1515 Central Avenue N.E.
Minneapolis, Minnesota 55413

Aluminum Company of America
794 Alcoa Building
Pittsburgh, Pennsylvania 15219

Amchem Products, Incorporated
Agr. Chemicals Division
Ambler, Pennsylvania 19002

American Air Filter Co., Inc.
215 Central Avenue
Louisville, Kentucky 40208

American Bakers Assoc.
20 N. Wacker Drive
Chicago, Illinois 60606

American Brahman Breeders Assn.
4815 Gulf Freeway
Houston, Texas 77023

American Cancer Society, Inc.
219 E. 42nd Street
New York, New York 10017

American Chain & Cable Co., Inc.
Box 430 Bridgeport
Bridgeport, Connecticut 06602

American Foot Care Institute, The
1775 Broadway
New York, New York 10019

American Gas Assn.
605 Third Avenue
New York, New York 10016

American Hardboard Assn.
20 N. Wacker Drive
Chicago, Illinois 60606

American Heart Association
44 E. 23rd Street
New York, New York 10010

American Institute of Architects
1735 New York Avenue, N.W.
Washington, D.C. 20006

American Institute of Cooperation
1200 17th Street, N.W.
Washington, D.C. 20036

American Institute of Steel
  Construction
101 Park Avenue
New York, New York 10017

American Music Conference
332 S. Michigan Avenue
Chicago, Illinois 60604

American Ostopathic Assn.
212 E. Ohio Street
Chicago, Illinois 60611

American Petroleum Inst.
1271 Avenue of the Americas
New York, New York 10020

American Plywood Assn.
Tacoma, Washington 98401

American Podiatry Assn.
3301 16th Street, N.W.
Washington, D.C. 20010

American Potash Inst., Inc.
1102 16th Street, N.W.
Washington, D.C. 20036

American Red Cross
150 Amsterdam Avenue
New York, New York 10023

American Society for Metals
Metals Park,
Ohio 44073

Am-Stand. Plumbing &
  Heating Div.
16 Johns Street
Princeton, New Jersey 08540

American Trucking Assns., Inc.
1616 P Street, N.W.
Washington, D.C. 20036

American Waterways Operators,
  Inc., The
1250 Connecticut Avenue
Washington, D.C. 20036

Ames Company, Inc.
819 McNaughton Street
Elkhart, Indiana 46514

AMPCO Metal, Inc.
Box 2004
Milwaukee, Wisconsin 53201

Anchor Serum Co.
2621 N. Belt Highway
St. Joseph, Missouri 64502

AP Parts Corp.
1801 Speilbusch Avenue
Toledo, Ohio 43601

Arbogast Co., Inc., Fred
313 W. North Street
Akron, Ohio 44303

Arcos Corp.
1500 S. 50th Street
Philadelphia, Pennsylvania 19143

ARMCO Steel Corp.
703 Curtis Street
Middletown, Ohio 45042

Armour and Co.
Box 9222
Chicago, Illinois 60690

Armstrong Cork Co.
Liberty Street
Lancaster, Pennsylvania 17604

Association Films, Inc.
600 Madison Avenue
New York, New York 10022

Atomics International Div.
Box 309
Canoga Park, California 91304

Austrian Institute
11 E. 52nd Street
New York, New York 10022

Automatic Lift Truck
101 W. 87th Street
Chicago, Illinois 60620

Babcock & Wilcox Co.
161 E. 42nd Street
New York, New York 10017

Babson Bros.
Dairy Research Service
2843 W. 19th Street
Chicago, Illinois 60623

Barber-Green Co.
Aurora, Illinois 60507

Barber Products, Inc.
2130 S. Bellaire Street
Denver, Colorado 80222

Bausch & Lomb, Inc.
635 St. Paul Street
Rochester, New York 14602

Bay State Film Prod., Inc.
Box 129
Springfield, Massachusetts 01001

Becton, Dickinson & Co.
Rutherford, New Jersey 07070

Behlen Mfg. Co.
Box 569
Columbus, Nebraska 68601

Bell & Howell Co.
7100 McCormick Road
Chicago, Illinois 60645

Bellingrath Gardens & Home
Theodore, Alabama 36582

Bell System Telephone Offices
150 West Street
New York, New York 10007

Bendix Corp., The
Sidney, New York 13838

Bergen Motion Picture Service
Route #46
Lodi, New Jersey 07644

Bird and Son, Inc.,
East Walpole, Massachusetts
02032

Blodgett Co., Inc.
50 Lakeside Avenue
Burlington, Vermont 05402

Boeing Company, The
Seattle, Washington 98124

Bonneville Power Administration
Box 3621
Portland, Oregon 97208

Bosch, Robert, Corp.
111 Crossways Park W.
Woodbury, New York 11797

Boyd Film Co.
1569 Selby Avenue
St. Paul, Minnesota 55104

Bray Studios, Inc.
729 Seventh Avenue
New York, New York 10019

Brett-Guard Corp.
160 Woodbine Street
Bergenfield, New Jersey 07621

Brooks Institute of Photography
2190 Alston Road
Santa Barbara, California 93103

Brown & Sharpe Mfg. Co.
Precision Park
N. Kingstown, Rhode Island
  02852

Brown Company
Berlin, New Hampshire 03570

Brown Shoe Co.
8300 Maryland Avenue
St. Louis, Missouri 63105

Bruce Co., Inc. W. L.
Box 12187 Memphis
Memphis, Tennessee 38112

Brunswick Corp.
623 S. Wabash Avenue
Chicago, Illinois 60605

Bryant Chucking Grinder Co.
Springfield, Vermont 05156

Bucyrus-Erie Co.
Box 56
South Milwaukee, Wisconsin
  53172

Buehler Ltd.
2120 Greenwood Street
Evanston, Illinois 60204

Bureau of Communication
  Research
267 W. 25th Street
New York, New York 10001

Bureau of Land Management
Washington, D.C. 20240

Bureau of Mines
4800 Forbes Avenue
Pittsburgh, Pennsylvania 15213

Bureau of Reclamation
Denver Federal Center
Denver, Colorado 80225

Bureau of Public Roads
Washington, D.C. 20235

Bureau of Sport Fisheries
Post Office & Federal Bldg. 801
Boston, Massachusetts 02109

California Mission Trails Assn.
25 W. Anapamu Street
Santa Barbara, California 93104

California Redwood Assn.
617 Montgomery Street
San Francisco, California 94111

Campbell Films
Saxtons River, Vermont 05154

Carey Salt Co.
1800 Carey Boulevard
Hutchinson, Kansas 67501

Cargill, Incorporated
Cargill Building
Minneapolis, Minnesota 55402

Central Soya Co.
Decatur, Indiana 46733

Cessna Aircraft Co.
5800 E. Pawnee Road
Wichita, Kansas 67201

Champion Spark Plug Co.
900 Upton Avenue
Toledo, Ohio 43601

Channel Master Corp.
Ellenville, New York 12428

Charleston Trident C. of C.
Box 975
Charleston, South Carolina 29402

Chemical Specialties Mfgrs. Assn.
50 E. 41 Street
New York, New York 10017

Chevron Chemical Co.
200 Bush Street
San Francisco, California 94120

Christopher-Gerard & Assocs.
4094 Fairlane Drive
Birmingham, Michigan 48010

Cincinnati Shaper Co.
Box 111
Cincinnati, Ohio 45211

Clark Equipment Co.
Battle Creek, Michigan 49016

Cleveland Range Co., The
971 E. 63rd Street
Cleveland, Ohio 44103

Cleveland Twist Drill Co.
Box 6656
Cleveland, Ohio 44101

Clevite Corp.
17000 St. Clair Avenue
Cleveland, Ohio 44110

Coleman Co., Inc.
250 N. St. Francis
Wichita, Kansas 67201

College Entrance Examination
 Board
475 Riverside Drive
New York, New York 10027

Collins, Miller & Hutchings, Inc.
333 W. Lake Street
Chicago, Illinois 60606

Colonial Broach & Machinery Co.
21601 Hoover Road
Warren, Michigan 48089

Colorado School of Mines
Golden, Colorado 80401

Columbus Pars
1801 Spielbusch Avenue
Toledo, Ohio 43601

Commercial Museum
34 St. & Convent Avenue
Philadelphia, Pennsylvania 19104

Commonwealth of Massachusetts
182 Tremont Street
Boston, Massachusetts 02111

Compressed Air & Gas Inst.
55 Public Square
Cleveland, Ohio 44113

Consumers Union Film Library
267 W. 25th Street
New York, New York 10001

Copley Products
434 Downer Place
Aurora, Illinois 60506

Creativision, Inc.
1780 Broadway
New York, New York 10019

Crosby Aeromarine Co.
5203 W. Highway 98
Panama City, Florida 32401

Crosby Forest Products Co.
Lock Drive 71
Picayune, Mississippi 39466

Daisy Manufacturing Co.
Rogers, Arkansas 72756

Dana Parts Co.
Box 500
Hagerstown, Indiana 47346

Dartmouth College Films
Hanover, New Hampshire 03755

Davey Tree Expert Co.
Kent, Ohio 44240

Detroit Broach & Machinery Co.
950 S. Rochester Road
Rochester, Michigan 48063

DeVilbiss Co., The
300 Phillips Avenue
Toledo, Ohio 43601

De-Arco
510 Eighth Avenue
Lake City, Minnesota 55041

Diddle Glaser, Inc.
1200-1500 Graphic Arts Road
Emporia, Kansas 66801

DoAll Co.
254 N. Laurel Avenue
Des Plaines, Illinois 60016

DuPont De Nemours & Co., Inc.
Wilmington, Delaware 19898

Eastern Air Lines, Inc.
Film Lab
International Airport
Miami, Florida 33159

Eastern Express, Inc.
1450 Wabash Avenue
Terre Haute, Indiana 47808

Eastman Chemical Products, Inc.
Kingsport, Tennessee 37662

Eastman Kodak Co.
343 State Street
Rochester, New York 14650

Eaton Yale & Towne, Inc.
11000 Roosevelt Boulevard
Philadelphia, Pennsylvania 19115

Eberle Tanning Co.
Westfield, Pennsylvania 16950

Educational Development Labs, Inc.
284 E. Pulaski Road
Huntington, New York 11744

Electric Arc, Inc.
4 Saddle Road
Cedar Knolls, New Jersey 07927

Electric Storage Battery Co.
212 E. Eashington Avenue
Madison, Wisconsin 53703

Employers Mutual of Wausau
Wausau, Wisconsin 54402

Encyclopaedia Britannica Educational Corp.
425 N. Michigan Avenue
Chicago, Illinois 60611

Environmental Science Services Administration
Rockville, Maryland 20852

Ethicon, Inc.
Route 22
Somerville, New Jersey 08876

Eutectic Welding Alloys Corp.
40-40 172nd Street
Flushing, New York 11358

Ex-Cell-O Corporation
Detroit, Michigan 48232

Farm Film Foundation
1425 H Street, N.W.
Washington, D.C. 20005

Federal Products Corp.
1144 Eddy Street
Providence, Rhode Island 02901

Federation of Societies for Paint Technology
121 S. Broad Street
Philadelphia, Pennsylvania 19107

Fellows Gear Shaper Co.
78 River Street
Springfield, Vermont 05156

Fidelity Advertising
Box 1739
Abilene, Texas 79604

Films of the Nations
  Distributors, Inc.
5113-16 Avenue
Brooklyn, New York 11204

First Natl. Bank of Minneapolis
120 S. 6th Street
Minneapolis, Minnesota 55402

FMC Corp.
1617 J.F.K. Boulevard
Philadelphia, Pennsylvania 19103

Food & Agriculture Organization
  of the U.N.
1325 C St. N.W.
Washington, D.C. 20437

Ford Motor Co.
Film Library
American Road
Dearborn, Michigan 48121

Freeport Sulphur Co.
161 E. 42nd Street
New York, New York 10017

French Co., R. T.
1 Mustard Street
Rochester, New York 14609

Frigidaire Division
300 Taylor Street
Dayton, Ohio 45401

Gardner-Denver Co.
Quincy, Illinois 62302

Gehl Bros. Mfg. Co.
143 Water Street
West Bend, Wisconsin 53095

General Dynamics Corp.
Box 748
Fort Worth, Texas 76101

General Mills, Inc.
9200 Wayzata Boulevard
Minneapolis, Minnesota 55440

General Motors Corp.
1775 Broadway
New York, New York 10019

General Telephone Co.
P. O. Box 889
Santa Monica, California 90406

Geometric Toll Co. Div.
1 Valley Street
New Haven, Connecticut 06515

Ghana Information Services
565 Fifth Avenue
New York, New York 10017

Glass Container Manufactures
  Institute, Inc.
330 Madison Avenue
New York, New York 10017

Glenwood Springs C. of C.
Box 97
Glenwood Springs, Colorado
  81601

Goodrich Co., The B. F.
500 S. Main Street
Akron, Ohio 44318

Goodyear Tire & Rubber Co., The
Akron, Ohio 44316

Gorton Machine Corp.
Box 705
Racine, Wisconsin 53401

Greenlee Bros. & Co.
2136 12th Street
Rockford, Illinois 61101

Grinding Wheel Institute
2130 Keith Building
Cleveland, Ohio 44115

Guidance Information Center
Academy Avenue
Saxtons River, Vermont 05154

Gulf Eastern Film Library
600 Grand Avenue
Ridgefield, New Jersey 07657

Gypsum Assn.
201 N. Wells Street
Chicago, Illinois 60605

Hammermill Paper Co.
1557 E. Lake Road
Erie, Pennsylvania 16512

Hanna Coal Co.
Cadiz, Ohio 43907

Harley-Davidson Motor Co.
3700 W. Juneau Avenue
Milwaukee, Wisconsin 53201

Harris-Seybold Co.
4510 E. 71st Street
Cleveland, Ohio 44105

Harshe-Rotman & Druck, Inc.
108 N. State Street
Chicago, Illinois 60602

Harvest Films, Inc.
25 W. 43rd Street
New York, New York 10036

Hear Foundation
301 E. Del Mar Blvd.
Pasadena, California 91106

Heinz Co., H. J.
Box 57
Pittsburgh, Pennsylvania 15230

Henry Ford Museum &
  Greenfield Village
Dearborn, Michigan 48124

Hercules Incorporated
910 Market Street
Wilmington, Delaware 19899

High Standard Sporting Firearms
1817 Dixwell Avenue
Hamden, Connecticut 06514

Hoober, Inc., J. M.
Union Stock Yds.
Lancaster, Pennsylvania 17604

Horan Engraving Co., Inc.
44 W. 28th Street
New York, New York 10001

Hughes Aircraft Co.
Box 90515
Los Angeles, California 90009

Huntington Labs, Inc.
Huntington, Indiana 46750

Ideal Cement Co.
821 17th Street
Denver, Colorado 80202

Ideal Pictures, Inc.
1919 Church Street
Evanston, Illinois 60201

Illinois Central R. R.
135 E. 11th Place
Chicago, Illinois 60605

Imperial-Eastman Corp.
6300 W. Howard Street
Chicago, Illinois 60648

Imperial Sugar Co.
Sugar Land, Texas 77478

Industrial Tectonics, Inc.
Box 1128
Ann Arbor, Michigan 48106

Ingersoll-Rand Co.
Phillipsburg, New Jersey 08865

Institute of Makers of Explosives
420 Lexington Avenue
New York, New York 10017

Institute of Scrap Iron & Steel
1729 H Street, N.W.
Washington, D.C. 20006

Insulation Board Institute
205 W. Touhy
Park Ridge, Illinois 60068

Insurance Information Institute
267 W. 25th Street
New York, New York 10001

International Association of Electro-Typers & Stereo
758 Leader Building
Cleveland, Ohio 44114

International Association of Machinists
909 Machinists Building
Washington, D.C. 20036

International Business Machine Corp.
330 Madison Avenue
New York, New York 10017

International Business Machine Corp.
Lexington, Kentucky 40507

International Minerals & Chemical Corp.
Box 71
Carlsbad, New Mexico 88220

John Hancock Mutual Life Insurance Co.
200 Berkeley Street
Boston, Massachusetts 02117

Johns-Manville Sales Corp.
22 E. 40th Street
New York, New York 10016

Johnson & Johnson
New Brunswick, New Jersey 08903

John Wood Co.
Broadway & Wood Streets
Muskegon, Michigan 49444

Jones & Lamson
160 Clinton Street
Springfield, Vermont 05156

Kaiser Aluminum & Chemical Corp.
300 Lakeside Drive
Oakland, California 94604

Kaiser Steel Corp.
300 Lakeside Drive
Oakland, California 94604

Kearney & Trecker Corp.
11000 Theo. Trecker Way
Milwaukee, Wisconsin 53214

Kendall Refining Co.
77 N. Kendall Avenue
Bradford, Pennsylvania 16701

Kerr-McGee Oil Corp.
Kerr-McGee Building
Oklahoma City, Oklahoma 73101

Keystone Steel & Wire Co.
Peoria, Illinois 61607

Kiekhaefer Corp.
1939 Pioneer Road
Fond du Lac, Wisconsin 54935

Kimberly-Clark Corp.
Neenah, Wisconsin 54956

Kirk & Son, Inc., Samuel
Kirk Avenue
Baltimore, Maryland 21218

Korean Information Office
1827 Jefferson Place, N.W.
Washington, D.C. 20036

Leather Industries of America, Inc.
411 Fifth Avenue
New York, New York 10016

Lederle Laboratories
Pearl River, New York 10965

Leeds & Northrup Co.
4901 Stenton Avenue
Philadelphia, Pennsylvania 19144

Lewis, Inc., Bernard L.
Empire State Building
New York, New York 10001

Libbey-Owens-Ford Glass Co.
811 Madison Street
Toledo, Ohio 43624

Lilly and Co., Eli
Indianapolis, Indiana 46206

Lima Register Co.
1790 N. Cable Road
Lima, Ohio 45802

Lincoln Electric Co., The
Cleveland, Ohio 44117

Lockheed-Georgia Co.
Zone 30, B-2 Building
Marietta, Georgia 30061

McLouth Steel Corp.
300 S. Livernois Street
Detroit, Michigan 48217

Macwhyte Wire Rope Co.
Kenosha, Wisconsin 53140

Mann Packing Co., Inc.
Box 908
Salinas, California 93901

Marathon Oil Co.
539 S. Main Street
Findlay, Ohio 45840

Market Forge Co.
Everett, Massachusetts 02149

Mason & Dixon Lines, Inc., The
Kingsport, Tennessee 37662

Maynard Research Council, Inc.
718 Wallace Avenue
Pittsburgh, Pennsylvania 15221

Maytag Co., The
Newton, Iowa 50208

Meehanite Metal Corp.
New King Street
White Plains, New York 10604

Merck Sharp & Dohme
West Point, Pennsylvania 19486

Merrell Co., The William S.
Cincinnati, Ohio 45215

Metropolitan Life Insurance Co.
1 Madison Avenue
New York, New York 10010

Michigan Consolidated Gas Co.
1 Woodward Avenue
Detroit, Michigan 48226

Micromatic Hone Corp.
8100 Schoolcraft Avenue
Detroit, Michigan 48238

Midland Cooperatives, Inc.
739 Johnson Street, N.E.
Minneapolis, Minnesota 55413

Miehle Co., The
2011 Hastings Street
Chicago, Illinois 60608

Mine Safety Appliances Co.
201 N. Braddock Avenue
Pittsburgh, Pennsylvania 15208

Miniature Precision Bearings, Inc.
Precision Park
Keene, New Hampshire 03431

Modern Equipment Co.
Box 266
Port Washington, Wisconsin
53074

Mokin Productions, Inc., Arthur
17 W. 60th Street
New York, New York 10023

Monarch Machine Tool Co.
Sidney, Ohio 45365

Montana Aeronautics Commission
Box 1698
Helena, Montana 59601

Mueller Brass Co.
Port Huron, Michigan 48061

NABAC
Box 500
Park Ridge, Illinois 60068

NAC Association
1155 15th Street, N.W.
Washington, D.C. 20005

National Aeronautics & Space
  Administration
Code FAD-2
Washington, D.C. 20546

National Association of
  American Business Clubs
Box 5127
High Point, North Carolina 27262

National Association of
  Homebuilders, The
1625 L Street, N.W.
Washington, D.C. 20036

National Association of
  Manufacturers
277 Park Avenue
New York, New York 10017

National Association of Mutual
  Insurance Companies
2166 E. 46th Street
Indianapolis, Indiana 46205

National Association of Plumbing
  Heat Cooling Contractors
1016 20 St., N.W.
Washington, D.C. 20036

National Bureau of Standards
Washington, D.C. 20234

National Committee Careers in
  Medical Technology
9650 Rockville Pike
Washington, D.C. 20014

National Electrical Manufacturers
  Association
155 E. 44th Street
New York, New York 10016

National Gallery of Art
Washington, D.C. 20565

National Institute of Rug
  Cleaning, Inc.
1815 N. Ft. Myer Drive
Arlington, Virginia 22209

National Lubricating Grease
  Institute
4635 Wyandotte Street
Kansas City, Missouri 64112

National Medical AV Center
  (Annex)
Station K
Atlanta, Georgia 30324

National Paint, Varnish, &
  Lacquer Association, Inc.
1500 R. I. Avenue, N.W.
Washington, D.C. 20005

National Plant Food Inst.
1700 K Street, N.W.
Washington, D.C. 20006

New Holland Div. of
Sperry Rand Corp.
New Holland, Pennsylvania 17557

North American Aviation, Inc.
3370 Miraloma Avenue
Anaheim, California 92803

North Central Airlines, Inc.
6201 34 Avenue, South
Minneapolis, Minnesota 55450

North Dakota State Wheat
Commission
316 N. 5th
Bismarck, North Dakota 58502

Northrop Corp.
9744 Wilshire Boulevard
Beverly Hills, California 90212

Northwestern University
Evanston, Illinois 60201

Norton Company
Worcester, Massachusetts 01606

O'Brien Corp., The
2001 W. Washington
South Bend, Indiana 46621

Office of Civil Defense
Fort George G. Meade,
Maryland 20755

Oilgear Co., The
1560 W. Pierce Street
Milwaukee, Wisconsin 53204

Onsrud Machine Works, Inc.
7720 N. Lehigh Avenue
Niles, Illinois 60648

Optical Gaging Prods., Inc.
26 Forbes Street
Rochester, New York 14611

Oregon Drilling Assn.
Box 546
Salem, Oregon 97308

O'Sullivan Rubber Corp.
Box 603
Winchester, Virginia 22601

Owens-Corning Fiberglas Corp.
Natl. Bk. Building
Toledo, Ohio 43601

Pennsalt Chemicals Corp.
3 Penn Center
Philadelphia, Pennsylvania 19012

Perlite Institute, Inc.
45 W. 45th Street
New York, New York 10036

Personal Products, Inc.
Van Liew Avenue
Milltown, New Jersey 08850

Pfizer Medical Film Library
267 W. 25th Street
New York, New York 10001

Pharmaceutical Manufacturers
Assoc.
1155 15 Street, N.W.
Washington, D.C. 20005

Piper Aircraft Corp.
Lock Haven, Pennsylvania 17745

Post Office Dept.
8th Ave. & 33rd Street
New York, New York 10001

Pratt & Whitney Inc.
Charter Oak Boulevard
W. Hartford, Connecticut 06101

Proto Tool Co.
2209 Santa Fe Avenue
Los Angeles, California 90054

Pure Oil Co.
200 E. Golf Road
Palatine, Illinois 60067

Quincy Compressor Co.
217 Maine Street
Quincy, Illinois 62302

Radiant Films
220 W. 42 Street
New York, New York 10036

Radim Films
220 W. 42nd Street
New York, New York 10036

Ransburg Electro-Coating Corp.
3939 W. 56 Street
Indianapolis, Indiana

REA Express
219 E. 42nd Street
New York, New York 10017

Reichhold Chemicals, Inc.
525 N. Broadway
White Plains, New York 10602

Republic Steel Corp.
224 E. 131st Street
Cleveland, Ohio 44108

Reynolds Metals Co.
Box 2346
Richmond, Virginia 23218

Ringsby Truck Lines, Inc.
3201 Ringsby Court
Denver, Colorado 80216

Robins Co., Inc., A. H.
1407 Cummings Drive
Richmond, Virginia 23220

Rocketdyne Division
6633 Canoga Avenue
Canoga Park, California 91304

Rockford Machine Tool Co.
2500 Kishwaukee Street
Rockford, Illinois 61101

Rockwell Mfg. Co.
300 N. Lexington Street
Pittsburgh, Pennsylvania 15208

Rohm & Haas Co.
Huntsville, Alabama 35807

Ronson Corp.
Woodbridge, New Jersey 07905

Roquefort Association, Inc.
41 E. 42nd Street
New York, New York 10017

Roses, Inc.
217½ Ann Street
East Lansing, Michigan 48823

Rothacker, Inc.
241 W. 17th Street
New York, New York 10011

Rust-Oleum Corp.
2425 Oakton Street
Evanston, Illinois 60204

Rutledge Drilling Co.
Box 2303
Santa Fe, New Mexico 87501

Saginaw Steering Gear Div.
3900 Holland Road
Saginaw, Michigan 48605

St. John's College
Annapolis, Maryland 21404

St. Paul Films
7050 Pinehurst
Dearborn, Michigan 02146

Selmer, Inc., H. & A.
Box 310
Elkhart, Indiana 46514

Shell Oil Co.
450 N. Meridian
Indianapolis, Indiana 46204

Sierra Bullets, Inc.
10532 S. Painter Avenue
Santa Fe Springs, California 90670

Silken, Inc., Paul
21 W. 46th Street
New York, New York 10036

Sikorsky Aircraft
Stratford, Connecticut 06497

Simonds Saw & Steel Co.
470 Main Street
Fitchburg, Massachusetts 01420

Simplex Time Recorder Co.
S. Lincoln Street
Gardner, Massachusetts 01441

Simplot Co., J. R.
Box 912
Pocatello, Idaho 83201

Six Flags Over Texas
Box 191
Arlington, Texas 76010

Smith Corp., A. O.
Box 584
Milwaukee, Wisconsin 53201

Smithers-Oasis
Box 118
Kent, Ohio 44240

Smith, Kline & French Labs
1500 Spring Garden St.
Philadelphia, Pennsylvania 19101

Smithsonian Institute
Museum of History & Technology
Washington, D.C. 20560

Speer, Inc.
Box 641
Lewiston, Idaho 83501

Standard Pressed Steel Co.
Jenkintown, Pennsylvania 19046

Starrett Co., The L. S.
121 Crescent Street
Athol, Massachusetts 01331

State College at Bridgewater
Bridgewater, Massachusetts 02324

Sterling Movies, Inc.
43 W. 61st Street
New York, New York 10023

Stout State University
Menomonie, Wisconsin 54751

Structural Clay Products Inst.
1520 18th Street, N.W.
Washington, D.C. 20036

Superior Electric Co., The
383 Middle Street
Bristol, Connecticut 06010

Swank Motion Pictures, Inc.
201 S. Jefferson
St. Louis, Missouri 63130

Taft Youth & Adult Center
172 St. & Sheridan Avenue
Bronx, New York 10457

Tea Council Film Library
267 W. 25th Street
New York, New York 10016

Tektronix, Inc.
Box 500
Beaverton, Oregon 97005

Texaco Inc.
2100 Hunters Point Avenue
Long Island City, New York 11101

Thiokol Chemical Corp.
Box 27
Bristol, Pennsylvania 19007

Thiokol Chemical Corp.
Box 1296
Trenton, New Jersey 08607

3M Company
2501 Hudson Road
St. Paul, Minnesota 55119

Timken Roller Bearing Co., The
1835 Dueber Avenue, S.W.
Canton, Ohio 44706

Topps Chewing Gum, Inc.
254 36th Street
Brooklyn, New York 11232

Tribune Films, Inc.
38 West 32nd Street
New York, New York 10001

TRW Systems
1 Space Park
Redondo Beach, California 90278

Tulare County C. of C.
Court House
Visalia, California 93278

Underwriters' Labs, Inc.
207 E. Ohio Street
Chicago, Illinois 60611

Union Carbide Corp.
270 Park Avenue
New York, New York 10017

Union Fork & Hoe Co.
500 Dublin Avenue
Columbus, Ohio 43215

Union Pacific Railraod
1416 Dodge Street
Omaha, Nebraska 68102

Union Stock Yard & Transit Co.
Union Stock Yards
Chicago, Illinois 60609

United Aircraft Corp.
400 Main Street
E. Hartford, Connecticut 06108

United Gas Corp.
Box 1407
Shreveport, Louisiana 71102

U.S. Army Engineer Div.
1114 Commerce Street
Dallas, Texas 75202

U.S. Atomic Energy Commission
376 Hudson Street
New York, New York 10014

U.S. Baird Corp.
1700 Stratford Avenue
Stratford, Connecticut 06497

U.S. Borax & Chemical Corp.
3075 Wilshire Boulevard
Los Angeles, California 90005

U.S. Bureau of Census
Washington, D.C. 20233

U.S. Electrical Motors
901 N. Broadway
White Plains, New York 10603

U.S. Geological Survey
Washington, D.C. 20242

U.S. Industries, Inc.
6499 W. 6th Street
Chicago, Illinois 60638

U.S. Plywood Corp.
777 Third Avenue
New York, New York 10017

U.S. Rubber Co.
Naugatuck, Connecticut 06771

U.S. Savings Bonds Division
Washington, D.C. 10017

U.S. Steel Corp.
71 Broadway
New York, New York 10006

U.S. Stoneware Co.
Akron, Ohio 44309

Univac Div., Sperry Rand
Box 8100
Philadelphia, Pennsylvania 19101

Universal Education & Visual Arts
221 Park Avenue South
New York, New York 10003

University of Idaho
Moscow, Idaho 83842

University of Michigan
Frieze Building
Ann Arbor, Michigan 48104

University of Mississippi, The
University, Mississippi 38677

University of North Carolina
Chapel Hill, North Carolina 27514

University of So. California
University Park
Los Angeles, California 90007

Univis, Inc.
Vision Park
Ft. Lauderdale, Florida 33310

University of Washington
Seattle, Washington 90105

Upjohn Co., The
7000 Portage Road
Kalamazoo, Michigan 49001

Vanadium Corp. of America
200 Park Avenue
New York, New York 10017

Vermiculite Institute
208 S. LaSalle Street
Chicago, Illinois 60604

Vickers, Inc.
Box 302
Troy, Michigan 48084

Vocational Rehabilitation Administration
Washington, D.C. 20201

VR/Wesson Co.
800 Market Street
Waukegon, Illinois 60086

Webb Co., Jervis B.
9000 Alpine Avenue
Detroit, Michigan 48204

Western Pacific Railraod
516 Fifth Avenue
New York, New York 10036

Wexler & Sporty, Inc.
125 Lafayette Street
New York, New York 10013

White Sands Missile Range
White Sands Missile Range, New Mexico 88002

Woodward Governor Co.
5001 N. 2nd Street
Rockford, Illinois 61101

World University Service
20 W. 40th Street
New York, New York 10018

Wurlitzer Co., The
DeKalb, Illinois 60115

Wurtele Film Prods.
Box 504
Orlando, Florida 32802

Wykoff Steel Div.
Box 29
Ambridge, Pennsylvania 15003

Wyeth Film Library
Box 8299
Philadelphia, Pennsylvania 19101

Yeshiva University
Amsterdam Ave. & 186th
 Street
New York, New York 10033

Youngstown Sheet & Tube Co.,
 The
Box 900
Youngstown, Ohio 44501

Zeiss, Inc., Carl
444 Fifth Avenue
New York, New York 10018

Appendix M

Sources of 8 and 16mm Motion-Picture and Slide-film Libraries—Sales and Rentals

The following film sources are generally more applicable to the teaching of chemistry and physics. Catalogues should be obtained from the sources located closest to the teacher. The state universities will, as a general rule, offer the most complete subject-matter oriented films, but the other sources should not be overlooked. This list is by no means complete, and the teacher should add the names of other sources in his state.

ALABAMA

Auburn University
Main Library
Auburn, Alabama 36830

A-V Film Service
2114 Eighth Avenue N.
Birmingham, Alabama 35203

Moffitt, John R., Company
Box 102
Montgomery, Alabama 36101

Photo-Sound, Incorporated
1043 Adams Avenue
Montgomery, Alabama 36104

ALASKA

Department of Health & Welfare
Box 3-2000
Juneau, Alaska 99801

Pictures, Incorporated
811 Eighth Avenue
Anchorage, Alaska 99501

University of Alaska
AV Communications
College, Alaska 99735

ARIZONA

Equipment Company
402 S. 23rd Avenue
Box 551
Phoenix, Arizona 85001

University of Arizona
AV Services
Tucson, Arizona 85721

ARKANSAS

Arkansas State College
AV Center
State College, Arkansas 72467

CALIFORNIA

Aims Instructional Media Service
Box 1010
Hollywood, California 90028

Alemann Films
P. O. Box 76244
Los Angeles, California 90005

American Documentary Films
379 Bay Street
San Francisco, California 94233

Atlantis Products, Incorporated
1232 La Granada Drive
Thousand Oaks, California 91360

Bailey Films
6509 De Longpre Avenue
Hollywood, California 90028

Better Selling Bureau
1150 W. Olive Avenue
Burbank, California 91506

Classroom Film Distributors
5610 Hollywood Boulevard
Hollywood, California 90028

Copley Productions
Box 1530
La Jolla, California 92038

Cottage Films
13061 Jefferson
Garden Grove, California 92541

Elkins, Herbert M., Company
10031 Commerce Avenue
Tujunga, California 91042

Family Films
5823 Santa Monica Blvd.
Hollywood, California 90038

Film Associates of California
11559 Santa Monica Boulevard
Los Angeles, California 90025

International Communications
 Films
1371 Reynolds Avenue
Santa Ana, California 92705

J-D Audio Visual
1700 E. Walnut Street
Pasadena, California 91106

Johnson Hunt Products
Box 96
Coronado, California 92118

Knight's Library, Incorporated
3911 Normal Street
San Diego, California 92104

Long Filmslide Service
7505 Fairmount Avenue
El Cerrito, California 94530

McCurry Companies
8th and Eye Streets
Box 838
Sacramento, California 95840

Merkle Film Library
4001 Atlantic
Long Beach, California 90807

Moody Institution of Science
12000 E. Washington Boulevard
Whittier, California 90606

Munday & Collins, Incorporated
270 Willow Street
Box 5669
San Jose, California 95150

Pacific Film Library
104 Fountain Avenue
Pacific Grove, California 93950

Pyramid Films
Box 1048
Santa Monica, California 90406

University of California
Extension Media Center
Berkeley, California 94720

COLORADO

Akin & Bagshaw Film Library
2027 E. Colfax Avenue
Denver, Colorado 80206

Colorado State College
Instructional Materials Center
Greeley, Colorado 80631

Cromars' AV Center
1200 Stout Street
Denver, Colorado 80204

Thorne Films, Incorporated
1229 University Avenue
Boulder, Colorado 80302

University of Colorado
Stadium 348
Boulder, Colorado 80302

Visual Aid Center, Incorporated
1457 S. Broadway
Denver, Colorado 80210

CONNECTICUT

Aetna Life & Casualty
Information & Educational
 Department
Hartford, Connecticut 06115

H-B Motion Picture Service,
 Incorporated
21 Audio Lane
New Haven, Connecticut 06519

Jay James Camera Shop,
 Incorporated
183 Fairfield Avenue
Bridgeport, Connecticut 06603

Rockwell Film & Projection
 Service
187 Allyn Street
Hartford, Connecticut 06103

University of Connecticut
AV Center
Storrs, Connecticut 06268

DISTRICT OF COLUMBIA

Bureau of Reclamation
19th & C Streets, N.W.
Washington, D.C. 20240

Department of State
Office of Media Services
Washington, D.C. 20520

Epperley, Roy G. & Company
725 12th Street, N.W.
Washington, D.C. 20005

Film Center
915 12th Street, N.W.
Washington, D.C. 20005

Smithsonian Institution,
AV Library
Washington, D.C. 20560

FLORIDA

A-V of Tampa, Incorporated
P. O. Box 10099
Tampa, Florida 33609

Florida State University
Educational Media Center
Tallahassee, Florida 32306

Imperial Film Company,
 Incorporated
2959 Edison Avenue
Jacksonville, Florida 32205

Photosound of Orlando
1020 N. Mills Avenue
Orlando, Florida 32803

University of Miami
Box 8021
Coral Gables, Florida 33124

World Color, Incorporated
Route 1
Ormond Beach, Florida 32074

GEORGIA

A-V Center
510 Walker Street
Augusta, Georgia 30902

Calhoun Company, Incorporated
121 Forrest Avenue
N. E. Atlanta, Georgia 30303

Hurn, F. L., Pictures
Box 1954
Savannah, Georgia 31402

Seban, Incorporated
430-11th Street
Columbus, Georgia 31902

University of Georgia
Film Library
Athens, Georgia 30601

HAWAII

Film Services of Hawaii,
  Incorporated
742 Ala Moana Boulevard
Honolulu, Hawaii 96813

University of Hawaii
AV Activities
Honolulu, Hawaii 96822

IDAHO

Boise College
Educational Film Library
Boise, Idaho 83706

Film Originals
Box 4072
Boise, Idaho 83705

Idaho State University
Educational Film Library
Pocatello, Idaho 83201

University of Idaho
AV Center
Moscow, Idaho 83843

ILLINOIS

Argonne National Laboratory
9700 S. Cass Avenue
Argonne, Illinois 60440

Audio Film Center
2138 E. 7th Street
Chicago, Illinois 60649

Coronet Instructional Films
65 E. S. Water Street
Chicago, Illinois 60601

Encyclopedia Britannica
Educational Corporation
425 N. Michigan
Chicago, Illinois 60611

Films Incorporated
1144 Wilmette Avenue
Wilmette, Illinois 60091

Illini AV Educational Service
1829 W. Main Street
Peoria, Illinois 61606

International Film Bureau,
  Incorporated
332 S. Michigan Avenue
Chicago, Illinois 60604

Lundgren's A-V
415 Seventh Street
Rockford, Illinois 61104

National Council Teachers of
  English
508 S. 6th Street
Champaign, Illinois 61820

Newenhouse, Henk
1825 Willow Road
Northfield, Illinois 60093

Northwestern University Film
  Library
828 Custer Avenue
Evanston, Illinois 60202

O'Hair Film Service
1443 N. 3rd Street
Springfield, Illinois 62702

Society for Visual Education, Incorporated
1345 Diversey Parkway
Chicago, Illinois 60614

South Suburban Film Coop.
400 Lakewood
Park Forest, Illinois 60466

Southern Illinois University
Learning Resources Service
Carbondale, Illinois 62901

Sportlite Films
20 N. Wacker Drive
Chicago, Illinois 60606

University of Illinois
704 S. 6th Street
Champaign, Illinois 61820

INDIANA

A-V Specialists
3753 Northrop Street
Fort Wayne, Indiana 46805

Burke's AV Center
2207 S. Michigan Street
South Bend, Indiana 46113

Gary Camera Company
619 Washington
Gary, Indiana 46402

Indiana State University
Stalker Hall
Terre Haute, Indiana 47808

Indiana University
AV Center
Bloomington, Indiana 47401

Purdue University
AV Center
Lafayette, Indiana 47097

Shoemaker Motion Picture Company
343 N. Capitol
Indianapolis, Indiana 46204

Smith & Butterfield
2800 Lynch Road
Evansville, Indiana 47711

Wayne Camera & Visual Equipment Company
1231 E. State Street
Ft. Wayne, Indiana 46805

IOWA

Iowa State University
Pearson Hall 121
Ames, Iowa 50010

Manbeck Pictures Corporation
3621A Wakonda Drive
Des Moines, Iowa 50321

Midwest Visual Education Service, Inc.
2204 Ingersoll
Des Moines, Iowa 50312

Pratt Sound Films, Incorporated
129 3rd Avenue
S. W. Cedar Rapids, Iowa 52404

State University of Iowa
Bureau of AV Instruction
Iowa City, Iowa 52240

KANSAS

Kansas State Department of Health
State Office Building
Topeka, Kansas 66612

Leffingwell's AV
109 N. 7th
Box 852
Salina, Kansas 67401

Lewis Film Service
1425 E. Central Avenue
Wichita, Kansas 67214

Smith, Steve, Cameras
623 Kansas Avenue
Box 1216
Topeka, Kansas 66601

University of Kansas
Bureau of Visual Instruction
Lawrence, Kansas 66044

KENTUCKY

AV Equipment Company,
  Incorporated
4511 Dixie Highway
Louisville, Kentucky 40212

Kentucky Department
  of Libraries
Box 537
Frankfort, Kentucky 40601

University of Kentucky
Taylor Education Building
Lexington, Kentucky 40506

LOUISIANA

Delta Pictures, Incorporated
327 Market
Shreveport, Louisiana 71101

Delta Visual Service, Incorporated
715 Girod Street
New Orleans, Louisiana 70130

Leonard AV Service
1303 Prytania
New Orleans, Louisiana 70113

Louisiana State University
Agriculture Extension Service
Baton Rouge, Louisiana 70803

MAINE

Headlight Film Service
111 Ocean Street
South Portland, Maine 04106

University of Maine
Stevens Hall South
Orono, Maine 04473

MARYLAND

Folkemer Photo Service
3610 Washington Boulevard
Baltimore, Maryland 21217

Kunz, Incorporated
207-209 E. Patapsco Avenue
Baltimore, Maryland 21225

Lewy Film & Sound Service
853 N. Eutaw Street
Baltimore, Maryland 21201

University of Maryland
Film Library
College Park, Maryland 20742

White & Leonard, Incorporated
Main & St. Peters Streets
Salisbury, Maryland 21801

White, Nelson, C.
Ideal Pictures
102 W. 26th Street
Baltimore, Maryland 21218

MASSACHUSETTS

Cinema, Incorporated
234 Clarendon Street
Boston, Massachusetts 02116

Ealing Corporation
2225 Massachusetts Avenue
Cambridge, Massachusetts 02140

Ginn & Company
Back Bay
P. O. Box 191
Boston, Massachusetts 02117

Wholesome Film Center,
  Incorporated
20 Melrose Street
Boston, Massachusetts 02116

MICHIGAN

AV Education
15920 Grand River
Detroit, Michigan 48227

Brice Sound & Electric,
  Incorporated
2401 S. Dort Highway
Flint, Michigan 48507

Central Michigan University
AV Services
Mt. Pleasant, Michigan 48858

Krums AV
35 E. Michigan
Battle Creek, Michigan 49014

Newman Visual Education,
  Incorporated
2023 Eastern Avenue
South East
Grand Rapids, Michigan 49507

Standard Film Service
Box 4863
Detroit, Michigan 48216

UAW Education Department
800 E. Jefferson Avenue
Detroit, Michigan 48214

University of Michigan
Frieze Building
Ann Arbor, Michigan 48103

Van's AV Center
524 E. Michigan Avenue
Lansing, Michigan 48933

Wayne State University
5448 Cass Avenue
Detroit, Michigan 48202

Western Michigan University
AV Center
Kalamazoo, Michigan 49001

MINNESOTA

AV Center & South's Film Library
6422 W. Lake Street
Minneapolis, Minnesota 55426

Augsburg Publishing House
426 S. 5th Street
Minneapolis, Minnesota 55414

Hart AV Center
26 N. 1st Avenue
E. Duluth, Minnesota 55802

Knight Film Company
1114 Nicollet Avenue
Minneapolis, Minnesota 55403

University of Minnesota
2037 University Avenue
S. E. Minneapolis, Minnesota
  55455

World Wide Pictures
1313 Hennepin Avenue
Minneapolis, Minnesota 55403

MISSISSIPPI

Herschel Smith Company
119 Roach Street
Jackson, Mississippi 39205

University of Mississippi
Educational Film Library
University, Mississippi 38677

MISSOURI

AV Services, Incorporated
709 Westport Road
Kansas City, Missouri 64111

Cor-Rell Communications
  Company
5316 Pershing Avenue
St. Louis, Missouri 63112

Horizon Products
301 W. 73rd Street
Kansas City, Missouri 64114

Missouri Department of Public
  Health and Welfare
Jefferson City, Missouri 65101

Raymer-Johnson, Incorporated
1907 S. Kings Highway
St. Louis, Missouri 63110

Swank Motion Pictures,
  Incorporated
201 S. Jefferson Avenue
St. Louis, Missouri 63103

University of Missouri
AV Library
Columbia, Missouri 65201

Wingo's Camera Center
420 Main Street
Joplin, Missouri 64801

NEBRASKA

Kretschmer, J. G. & Company
Box 3301
Omaha, Nebraska 68103

Modern Sound Pictures, Inc.
1410 Howard Street
Omaha, Nebraska 68102

University of Nebraska
Bureau of AV Instruction
Lincoln, Nebraska 68508

NEVADA

Chism, Harold F.
School Supplies
638 W. 5th Street
Reno, Nevada 89503

University of Nevada
AV Communications Center
Reno, Nevada 89507

NEW HAMPSHIRE

Dartmouth College Films
Fairbanks Hall
Hanover, New Hampshire 03755

University of New Hampshire
AV Department
Durham, New Hampshire 03824

NEW JERSEY

Communication Arts,
  Incorporated
Box 478 Bernardsville
Bernardsville, New Jersey 07924

Department of Education
State House Annes
Trenton, New Jersey 08652

Film Presentation Company, Inc.
971 Bergen Street
Newark, New Jersey 07112

NEW MEXICO

University of New Mexico
AV Aids
Albuquerque, New Mexico 87106

NEW YORK

Alden Films
5113 16th Avenue
Brooklyn, New York 11204

American Documentary Films
336 W. 84th Street
New York, New York 10024

American Film Products
1540 Broadway
New York, New York 10036

American Heart Association
44 E. 23rd Street
New York, New York 10010

American Petroleum Institute
1271 Avenue of Americas
New York, New York 10020

Association Films, Incorporated
600 Madison Avenue
New York, New York 10022

Association Instructional
  Materials
600 Madison Avenue
New York, New York 10022

Audio Film Center/Ideal Pictures
34 MacQuesten Parkway South
Mt. Vernon, New York 10550

Audiovision Language
Technology Service
3508 Grand Central Station
New York, New York 10017

Australian News & Information
  Bureau
636 Fifth Avenue
New York, New York 10020

AV Instructionals
1142 Oak Street
Syracuse, New York 13203

Brandon Films, Incorporated
221 W. 57th Street
New York, New York 10019

Bray Studios, Incorporated
630 Ninth Avenue
New York, New York 10036

Brooklyn Union Gas Company
195 Montague Street
Brooklyn, New York 11201

Buchan Pictures
122 W. Chippewa Street
Buffalo, New York 14202

Business Educational Films
5113 16th
Brooklyn, New York 11204

Camera Center
596 Grand Street
Brooklyn, New York 11211

Carousel Films, Incorporated
Suite 1503
1501 Broadway
New York, New York 10036

Catholic M/P. Enterprises
2040 Chatterton Avenue
Bronx, New York 10473

CBS/HOLT Group
383 Madison Avenue
New York, New York 10017

Doubleday Multimedia
277 Park Avenue
New York, New York 10017

Du Art Film Labs
245 W. 55th Street
New York, New York 10019

Duncan, James E., Inc.
857 Portland Avenue
Rochester, New York 14603

Eye-Gate House, Incorporated
145-01 Archer
Jamaica, New York 11435

Films of the Nations
5113-16th Avenue
Brooklyn, New York 11204

Filmstrip House
432 Park Avenue South
New York, New York 10016

G. E. Educational Films
60 Washington Avenue
Schenectady, New York 12305

Hellinger, B., Company
40 E. 88th Street
New York, New York 10003

Instructional Cinema Service,
  Incorporated
29 E. 10th Street
New York, New York 10003

Life Filmstrips
Time & Life Building
New York, New York 10020

McGraw-Hill Films
1221 Avenue of Americas
New York, New York 10036

Museum of Modern Art
11 W. 53rd Street
New York, New York 10019

National Film Board of Canada
680 Fifth Avenue
New York, New York 10019

New York University Film Library
26 Washington Place
New York, New York 10003

Pathe Contemporary
330 W. 42nd Street
New York, New York 10036

Pathe News, Inc.
835 Broadway
New York, New York 10003

Pictocraft, Incorporated
145 Library Lane
Mamaroneck, New York 10543

Popular Science Audio Visuals
  Incorp.
355 Lexington Avenue
New York, New York 10017

Prism Enterprises, Incorporated
220 E. 23rd Street
New York, New York 10010

Renner Motion Picture Service,
  Incorp.
1609 Kenmore Avenue
Kenmore, New York 14217

Roach, Hal Studios
1501 Broadway
New York, New York 10036

Rogosin Films
144 Bleekcer Street
New York, New York 10012

Select Film Library Incorporated
115 W. 31st Street
New York, New York 10001

State University
Forest Extension
Syracuse, New York 13210

State University of New York
  at Buffalo
Communications Center
Buffalo, New York 14214

Syracuse University
1455 E. Colvin Street
Syracuse, New York 13210

The American Journal of Nursing
  Company
Film Library
267 W. 25th Street
New York, New York 10001

United Project Film Corporation
228 Franklin Street
Buffalo, New York 14202

Universal Educational & Visual
  Arts
221 Park Avenue South
New York, New York 10003

Visual Sciences
Suffern, New York 10901

Ward's Natural Science
  Establishment
300 Ridge Road
E. Rochester, New York 14622

Wayne County Library System
Mason & High Streets
Newark, New York 14513

Willoughby Peerless
415 Lexington Avenue
New York, New York 10017

Yeshiva University
526 W. 187th Street
New York, New York 10033

NORTH CAROLINA

National School & Ind.
  Corporation
14 Glenwood Avenue
Raleigh, North Carolina 27602

North Carolina State Board
  of Health
Film Library
Raleigh, North Carolina 27602

Tarmac AV Company
71 N. Market Street
Asheville, North Carolina 28801

University of North Carolina
Bureau of AV Education
Chapel Hill, North Carolina 27514

NORTH DAKOTA

North Dakota State Department
  of Health
Bismark, North Dakota 58501

North Dakota State University
Fargo, North Dakota 58102

OHIO

Academy Film Service,
  Incorporated
2110 Payne Avenue
Cleveland, Ohio 44114

Bartha Visual Education Service
1946 N. High Street
Columbus, Ohio 43201

Cousino Film Rental
1945 Franklin Avenue
Toledo, Ohio 43624

Eldridge AV Center
P. O. Box 11425
Columbus, Ohio 43225

Haile, Ralph V. & Associates
3524 Zumstein
Cincinnati, Ohio 45208

Harpster AV Equipment, Inc.
7777 Exchange Street
Cleveland, Ohio 44125

Kent State University
AV Center
Kent, Ohio 44240

Levy's Film & Project Service
1648 Pullan Avenue
Cincinnati, Ohio 45223

Mottas Films
1318 Ohio Avenue
N. E.
Canton, Ohio 44705

Ohio State University
Department Photo & Cinema
156 W. 19th Avenue
Columbus, Ohio 43210

Sunray Films, Incorporated
2005 Chester Avenue
Cleveland, Ohio 44114

Twyman Films, Incorporated
Box 605
Dayton, Ohio 45401

OKLAHOMA

Cory Motion Picture Equipment
522 N. Broadway
Oklahoma City, Oklahoma 73102

Holmes & Torbett
2902 Denver
Muskogee, Oklahoma 74401

Thompson Movie Supply,
   Incorporated
1740 S. Boston Avenue
Tulsa, Oklahoma 74119

University of Oklahoma
Educational Materials
Norman, Oklahoma 73069

Vaseco, Incorporated
Box 60274
Oklahoma City, Oklahoma 73106

OREGON

Moore's AV Center, Incorporated
234 S. E. 12th Avenue
Portland, Oregon 97214

Oregon State System of Higher
   Education
AV Instruction
Corvallis, Oregon 97331

PENNSYLVANIA

AV Center
14 Wood Street
Pittsburgh, Pennsylvania 15222

Clem Williams Films, Incorporated
2240 Noblestown Road
Pittsburgh, Pennsylvania

Curriculum Materials Corporation
1319 Vine Street
Philadelphia, Pennsylvania 19107

Curtis Audio-Visual Materials
Independence Square
Philadelphia, Pennsylvania 19105

Grise Film Library
901 French Street
Erie, Pennsylvania 16512

Hirt, Oscar H.
AV, Incorporated
41 N. 11th Street
Philadelphia, Pennsylvania 19107

International Film Center &
   AV Center, Incorp.
1915 Market Street
Philadelphia, Pennsylvania 19103

Lett, James, Company
Box 844
Harrisburg, Pennsylvania 17108

Lilley, J. P. & Sons, Incorporated
928 N. 3rd Street
Harrisburg, Pennsylvania 17105

Pennsylvania State University
AV Service
University Park, Pennsylvania
   16802

Psychological Cinema Register
AV Aids Library
University Park, Pennsylvania
   16802

Vath, L. C.
AV
449 N. Hermitage Road
Sharpsville, Pennsylvania 16150

Wespen AV Company
Hawthorn, Pennsylvania 16230

Your Lesson Plan
1319 Vine Street
Philadelphia, Pennsylvania 19107

RHODE ISLAND

Payne Motion Picture Service
20 High Street
Westerly, Rhode Island 02891

State Department of Education
Park & Hayes Street
Providence, Rhode Island 02908

SOUTH CAROLINA

University of South Carolina
AV Division
Columbia, South Carolina 29208

SOUTH DAKOTA

South Dakota State University
Film Library
Brookings, South Dakota 57006

Taylor Films
1009 Dakota Avenue, South
Huron, South Dakota 57250

TENNESSEE

Capitol Visual Aids
611 Dodds Avenue
Chattanooga, Tennessee 37404

Cokesbury
Box 801
Nashville, Tennessee 37202

Tennessee Visual Education
  Service
416A Broad Street
Nashville, Tennessee 37203

University of Tennessee
Film Service
Knoxville, Tennessee 37916

TEXAS

AV Services
2310 Austin Street
Houston, Texas 77004

Bauer Audio Video, Incorporated
2911 N. Haskell
Dallas, Texas 75204

Du Motion Picture Service
175-179 N. Cotton Street
El Paso, Texas 79901

Educational Materials Distributors
Box 9083
Austin, Texas 78756

E. & W. Sales Company
2020 Driskell
Beaumont, Texas 77706

Smith, Donald L., Company
1110 N. Main Avenue
San Antonio, Texas 78206

South Texas Visual, Incorporated
1917 Leopard Drive
9067
Corpus Christi, Texas 78401

Stevens Pictures of Texas
3019 Monticello
Dallas, Texas 76206

Stidham, Charles G., Company
1508 Fredericksburg Road
San Antonio, Texas 78201

Texas Education Association
Film Library
4006 Live Oak Street
Dallas, Texas 75204

Texas Educational Aids
4621 Fannin
Houston, Texas 77004

University of Texas
Visual Instruction
 Bureau
Main University
Austin, Texas 78712

Yocum's 8-16mm Film
 Headquarters
614 Du Pont Drive
Orange, Texas 77631

UTAH

Brigham Young University
Provo, Utah 84601

Deseret Book Company
44 E. So. Temple Street
Salt Lake City, Utah 84110

University of Utah
AV Bureau
Salt Lake City, Utah 84112

Utah State University
AV Aids
Logan, Utah 84321

Webb, George,
Sales Company
937 E. 33 South
Salt Lake City, Utah 84106

VERMONT

University of Vermont
Film Library & AV Service
Burlington, Vermont 05401

VIRGINIA

AV Center of Tidewater
135-137 E. Little Creek Road
Norfolk, Virginia 23505

Brand, Paul L. & Son
234 W. Broad Street
Falls Church, Virginia 22046

Brownings Ideal Pictures
200 E. Cary Street
Richmond, Virginia 23219

WASHINGTON

AV Center, Incorporated
1205 N. 45th Street
Seattle, Washington 98103

Library of Teaching Materials
Courthouse Annex
Kelso, Washington 98626

Rarig's Incorporated
2100 N. 45th Street
Seattle, Washington 98103

Superintendent of Public
 Instruction
Old Capitol Building
Olympia, Washington 98501

University of Washington
AV Services
Lewis Hall
Seattle, Washington 08105

WEST VIRGINIA

State Department of Health
Bureau of Public Health
Charleston, West Virginia 25305

West Virginia University
AV Library
Morgantown, West Virginia
 26506

WISCONSIN

Davis AV Company
210 S. Monroe
Green Bay, Wisconsin 54301

Milwaukee Public Museum
AV Center
Milwaukee, Wisconsin 53203

Roa's Films
1696 N. Astor Street
Milwaukee, Wisconsin 53202

Stout State College
AV Center
Menominee, Wisconsin 54751

University of Wisconsin
Bureau of AV Instruction
Madison, Wisconsin 53706

Wisconsin Sound Equipment
  Company, Inc.
4429 W. North Avenue
Milwaukee, Wisconsin 53208

WYOMING

University of Wyoming
AV Services
Laramie, Wyoming 82070